Systems

Volume 10

# Systems Thinking:
# Principles and Applications

Systems Engineering Program
Worcester Polytechnic Institute
100 Institute Road
Worcester, MA 01609
USA

Volume 7
Software Engineering in the Systems Context
Ivar Jacobson and Harold "Bud" Lawson, eds.

Volume 8
Using Systems Thinking to Solve Real-World Problems
Jamie P. Monat and Thomas F. Gannon

Volume 9
Creating, Analysing and Sustaining Smarter Cities. A Systems Perspective
Ian Abbott-Donnelly and Harold "Bud" Lawson, eds.

Volume 10
Systems Thinking: Principles and Applications
Jamie P. Monat and Matthew Amissah, eds.

Systems Series Editors
Jon P. Wade                              jon.wade@stevens.edu
Wolfgang Hofkirchner          wolfgang.hofkirchner@bcsss.org

# Systems Thinking:
# Principles and Applications

Edited by

## Jamie P. Monat

and

## Matthew Amissah

ISBN 978-1-84890-482-8


College Publications
Scientific Director: Dov Gabbay
Managing Director: Jane Spurr



http://www.collegepublications.co.uk

The Systems series publishes books related to Systems Science, Systems Thinking, Systems Engineering and Software Engineering that address trans-disciplinary Frontiers, Practice and Education

**Systems Science** having its contemporary roots in the first half of the 20th century is today made up of a diversity of approaches that have entered different fields of investigation. Systems Science explores how common features manifest in natural and social systems of varying complexity in order to provide scientific foundations for describing, understanding and designing systems.

**Systems Thinking** has grown during the latter part of the 20th century into highly useful discipline independent methods, languages and practices. Systems Thinking focuses upon applying concepts, principles, and paradigms in the analysis of the holistic structural and behavioral properties of complex systems – in particular the patterns of relationships that arise in the interactions of multiple systems.

**Systems and Software Engineering.** Systems Engineering has gained momentum during the latter part of the 20th century and has led to engineering related practices and standards that can be used in the life cycle management of complex systems. Software Engineering has continued to grow in importance as the software content of most complex systems has steadily increased and in many cases have become the dominant elements. Both Systems and Software Engineering focus upon transforming the need for a system into products and services that meet the need in an effective, reliable and cost effective manner. While there are similarities between Systems and Software Engineering, the unique properties of software often requires special expertise and approaches to life cycle management.

Systems Science, Systems Thinking, as well as Systems and Software Engineering can, and need to, be considered complementary in establishing the capability to individually and collectively "think" and "act" in terms of systems in order to face the complex challenges of modern systems.

This series is a cooperative enterprise between College Publications, the School of Systems and Enterprises at Stevens Institute of Technology and the Bertalanffy Centre for the Study of Systems Science (BCSSS).

For further information concerning the Systems Series see
http://www.collegepublications.co.uk/systems/

This book is a compendium of papers that were published between 2015 and 2024.

Dedicated to our friend and colleague, Tom Gannon

# Introduction

We wrote the following series of papers between 2015 and 2024. Some address the fundamental principles of Systems Thinking: what it is and what are its tools. Others address applications of Systems Thinking: how it may be beneficially applied to business, teaching, design, engineering, foreign policy, and nature. We hope that by assembling all these papers into one volume, we have developed a fuller picture of Systems Thinking.

That picture is still developing. Over the past 10 years, instead of the concepts of Systems Thinking coalescing into one agreed-upon paradigm, new interpretations continue to develop. Derek and Laura Cabrera, for example, have developed the DSRP approach. David Peter Stroh has applied Systems Thinking to social change. Many others have published works that attempt to put their own stamp on the definition of, tools for, and ways to apply Systems Thinking.

Our work has underscored the need to have a simple definition that is widely agreed-upon and easily understood by both experts in the field and laypeople. We have come up with this simple paradigm:

The Essence of Systems Thinking:

1. *Recognize* that you are dealing with a System, not an isolated issue.
2. Identify the components of the system and the system's boundaries.
3. *Understand* the Relationships among the components and the environment and how they drive system behavior.

The rest is all details.

# Table of Contents

# What Is Systems Thinking? A Review of Selected Literature Plus Recommendations

Monat, Jamie P., and Gannon, Thomas F.

*Reprinted from Am. J. of Systems Science, 4:2, 2015*

## Abstract

Systems Thinking is a popular current topic in the world of Systems Engineering. However, as yet there is no commonly accepted definition or understanding of it. In this paper, we analyze some of the popular Systems Thinking literature and attempt to identify common themes. We conclude that Systems Thinking is a perspective, a language, and a set of tools. Specifically, Systems Thinking is the opposite of linear thinking; holistic (integrative) versus analytic (dissective) thinking; recognizing that repeated events or patterns derive from systemic structures which, in turn, derive from mental models; recognizing that behaviors derive from structure; a focus on relationships vs components; and an appreciation of self-organization and emergence. Specific Systems Thinking tools include systemigrams, system archetypes, main chain infrastructures, causal loops with feedback and delays; stock and flow diagrams; behavior-over-time graphs, computer modeling of system dynamics, Interpretive Structural Modeling (ISM), and systemic root cause analysis.

## Introduction

Systems Thinking has its foundation in General Systems Theory (Bertalanffy) and has been applied to a wide range of fields and disciplines. It has great power in solving complex problems that are not solvable using conventional reductionist thinking. It can be used to explain dynamic non-linear behaviors like market reactions to new product introductions or predator-prey relationships; to understand complex socio-economic problems such as the effects of marijuana laws; and to understand the seemingly illogical behaviors of individuals, countries, and organizations such as ISIS's provocative actions.

1

However, many systems engineers do not fully grasp Systems Thinking—many believe it is simply the fundamental concepts of Systems Engineering as articulated by Kossiakoff et al. and Blanchard and Fabrycky, comprising V-diagrams, risk management, needs analysis, architecture and design, integration and test, and project management. This is not the case. Some practitioners have co-opted the term "Systems Thinking" to include *all* aspects of systems including general systems theory, cybernetics, family therapy, and Model-Based Systems Engineering (MBSE). We think this inappropriate. The *Systems Engineering Body of Knowledge* devotes a chapter to Systems Thinking. However, this chapter is a compendium of literature articles on systems thinking concepts, principles, and patterns. It is quite vague and does not appear to integrate the disparate articles into a cohesive whole. Furthermore, several key references (Meadows, Kim, Richmond) have been omitted. The dozens of books and articles written on Systems Thinking have some common threads, but different focuses and interpretations. In this paper we attempt to make sense of this chaos and develop a firm conceptual framework for Systems Thinking.

**Literature Review**

We do not purport to have done a comprehensive analysis of all systems thinking literature. We selected approximately 30 of the more popular works; works that we interpret to be "key" contributors to the understanding of systems thinking and that had "systems thinking" in either their title or subject description. To ensure that we had not missed any key references, we then submitted this list to 14 published experts in the field of Systems Thinking and asked for their suggestions regarding relevant literature; 9 of them (acknowledged at the end of this paper) were kind enough to reply with suggestions. We then evaluated their suggestions and added those that we believe advance the understanding of systems thinking (we did not include references focused on other aspects of systems, such as systems *engineering*, sources describing primarily predecessors to or precursors of systems thinking, or items that addressed sub-sub-elements of systems thinking such as the details of system dynamics programming.) The result is an edited list of approximately 33 references that we deem important to the understanding of systems thinking. These references are organized into 4 categories: Introductory Works, Applications of Systems Thinking, Self-organization and Emergence, and General Works.

*Introductory Works*

- Kim, Daniel H., *Introduction to Systems Thinking*. This 20-page booklet is probably the best concise introduction to systems thinking that is available. All of the basics are covered, including the definition of a system, the systems thinking "Iceberg Model," and systemic behavior including feedback loops and delays. The Iceberg Model argues that in a system, repeated events represent patterns and that patterns are invariably caused by systemic structure. In human-designed systems, systemic structure develops as a result of mental models. If one could read only one work to get a good grasp of Systems Thinking, this would be it.

- Richmond, Barry, *An Introduction to Systems Thinking with iThink*. iThink and Stella are 2 excellent system dynamics modeling software packages available from isee systems of Lebanon, NH. This instruction manual for these packages does much more than explain how to use the software; it is in fact a primer on systems thinking, covering such topics as system dynamics, feedback loops, stock-and-flow diagrams, main chain infrastructures, mental models, and non-linear effects. The most excellent aspect of this book, however, is its ability to relate everyday real-world situations to a systems thinking perspective. It is a terrific resource, whether one uses the software or not. Richmond includes a definition of Systems Thinking: "...systems thinking is the art and science of making reliable inferences about *behavior* by developing an increasingly deep understanding of the underlying *structure*."

- Meadows, Donella H., *Thinking in Systems: A Primer*. In our opinion, this is the seminal work on systems thinking. It was published posthumously from Dana Meadows's notes. It covers system definition, stock and flow diagrams,

3

feedback loops, resilience in systems, hierarchies, self-organization, unintended consequences, the 10 systemic archetypes, system leverage points, and rules for systems. However, it does a superb job in using real world examples (such as the inadvertent impact of DDT on bird eggshell thickness and the folly of spruce budworm control) to make its points. No student of systems thinking should miss this book.

- Anderson and Johnson, *Systems Thinking Basics, from Concepts to Causal Loops*. This relatively short book is a good study guide for introductory students of systems thinking. It is consistent with Kim, Meadows, and Richmond and covers the Iceberg Model, causal loop diagrams, archetypes, and behavior-over-time graphs. The book defines systems thinking as a set of tools (includes a "palette of systems thinking tools", a framework for looking at issues, and a language. It is somewhere between Kim and Meadows in its level of detail and examples.

- Kauffman, *Systems One: An Introduction to Systems Thinking*. This relatively early (1980) 40-page pamphlet is a concise introduction to the field. It discusses stability and feedback in systems, complexity, and archetypes, and it gives good examples of causal loop diagrams involving float valves, predator-prey relationships, thermostats, crime and punishment, compound interest, growth of power, and growth of knowledge. It is consistent with Meadows, Kim, Richmond, and Anderson and Johnson, but it does not discuss the Iceberg Model or dynamic modeling; it is thus less comprehensive than some other sources.

- Sweeney and Meadows, *The Systems Thinking Playbook*. This book attempts to teach many systems thinking principles through the use of games. It focuses on "habits of mind" --- identifying and then breaking them. Unfortunately, many of the games are sophomoric and don't make the points well. One exception is "Avalanche" in which several people try to lower a hula hoop simultaneously while supporting the hoop with just one finger. Contrary to everyone's mental model, the hoop goes up instead of down. This is an excellent demonstration of incorrect mental models and how they can dominate behavior. The book in general, however, falls short of teaching systems thinking principles via games.

- Galley, *Think Reliability: Investigation Basics—The Systems Approach*. This is a very good short article explaining how systems thinking enhances conventional root cause analysis. It argues that Systemic Root Cause Analysis should not identify a single root cause, but instead a root cause *system*: a paradigm, culture, environment, or set of attitudes that yield the specific identifiable causes. The sinking of the Titanic is used as an example.

- Aronson, *An Overview of Systems Thinking*, (*http://resources21.org/cl/files/project264_5674/OverviewSTarticle.pdf*). This article provides a good summary of how systems thinking is fundamentally different from reductionist thinking. It provides an example of how pesticides used to control insect damage to crops can give rise to alternative predatory insect damage that was not previously envisioned as an unintended consequence.

- Goodman, Kemeny, and Roberts, *The Language of Systems Thinking: 'Links' and 'Loops'*. This article provides a brief tutorial on the use of causal loops and delays to represent system behavior over time. It illustrates how they can be used to model the ups and downs of sales cycles, exponential growth or collapse in investment strategies, and stabilization in the number of patient visits to an outpatient clinic.

- Lawson, *A Journey Through the Systems Landscape*. The Lawson book discusses system classification and topologies as well as the Iceberg Model. It gets into systems archetypes, causal loop diagrams, system life cycles, and decision analysis, and it includes several good case studies involving crisis management, organizational development, architectural concepts, and ontology life cycle management. It is a good book and consistent with Meadows, Kim, and Richmond.

- Weinberg, *An Introduction to General Systems Thinking.* Weinberg presents several interesting and useful systems thinking concepts; among them the following plot (Figure 1) of Randomness versus Complexity, showing where systems (organized complexity) fit (a surprisingly large area of the plot.) (The original concepts of simplicity, organized complexity, and unorganized complexity are attributable to Weaver (1948)).

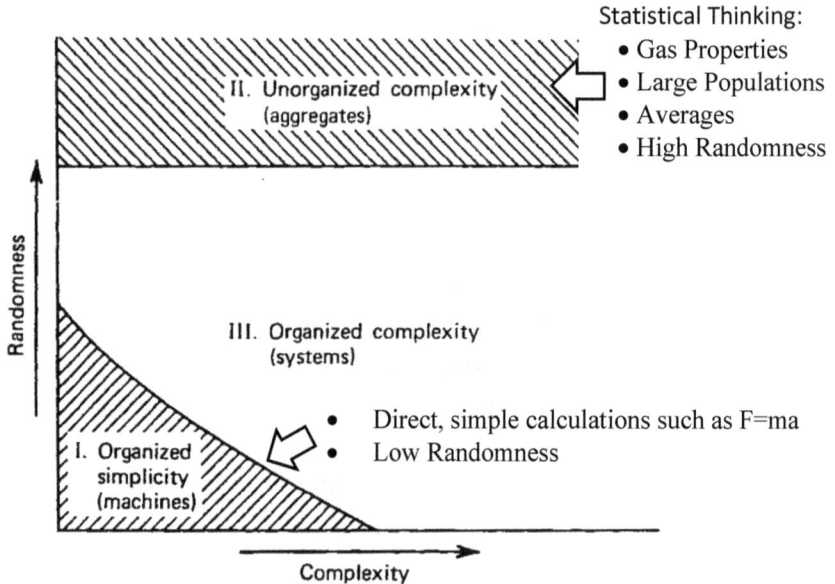

Figure 1. Weinberg's Systems Map of Randomness versus Complexity

Weinberg also proposes 3 Great Systems Thinking Questions:

- Why do I see what I see?
- Why do things stay the same?
- Why do things change?

He includes several good examples such as the inadvertent impacts of waste heat from nuclear reactors, the unintended consequences of targeted pesticides, and the detrimental effects of agricultural herbicides on fertility.

- Boardman and Sauser, *Systems Thinking: Coping With 21ˢᵗ Century Problems.* The best aspect of this book is its use of good, interesting, current systems thinking examples such as the impact on a rural community of new baby-boomer retirees, the September 11ᵗʰ attack on the world trade center, President Kennedy's national challenge to land men on the moon before 1970, and the privatization of the U.K. railroad industry. It also explains Systemigrams in good detail (see "Systems Thinking Tools"). The book refers extensively to other systems books such as Senge and Meadows. However, it is poorly organized and does not seem to provide a coherent, operational definition of systems thinking.

- Haines, *The Systems Thinking Approach.* Haines does a good job of applying systems thinking to business and discusses it in terms of current versus future states and how to move from one to the other. He argues for a focus on outcomes instead of activities and on processes and structures. While these are all elements of systems thinking, the book seems disjointed and does not provide a comprehensive, coherent picture of systems thinking.

- Gharajedaghi, *Systems Thinking.* Gharajedaghi's book argues that there are 5 systems principles: Openness, Purposefulness, Multi-dimensionality, Emergent Properties, and Counter-intuitiveness. He talks about System Context in terms of the environment, control, and influence; and he devotes substantial space to a health systems case study. However, the book is disorganized and does not present a clear explanation of systems thinking.

- Senge, *The Fifth Discipline.* Senge's 1990 book may not have started the trend in systems thinking, but it certainly accelerated it. In it, he provides some excellent examples of compensating feedback, application of systems thinking to terrorism, and systems archetypes. He also provides a generic definition: "Systems thinking is a discipline for seeing wholes. It is a framework for seeing interrelationships rather than things, for seeing patterns of change rather than "snapshots."... It is also a set of specific tools and techniques, originating in two threads: in "feedback" concepts of cybernetics and in "servo-mechanism" engineering theory..." Senge's book is pivotal because it applies systems thinking to management in organizations, and for that reason alone it is worth reading. Systems Thinking is (of course) the 5th discipline of a learning organization, the other 4 being personal mastery, mental models, building shared vision, and team learning. Senge argues that Systems Thinking is the most important because it integrates the other 4.

- Senge, Kleiner, Roberts, Ross, and Smith, *The Fifth Discipline Fieldbook*. While *The Fifth Discipline* presents the background and theory of systems thinking, this field book is much more applied. It presents many relevant examples and case studies (e.g. Sears's auto repair quality, the impact of Toyota's manufacturing quality on America's automobile expectations, water supply failures and fixes in Africa, and dealing with price wars) along with a systems thinking problem-solving approach. It discusses archetypes, systemic root cause analysis, system dynamics, and it provides a clear explanation of the Iceberg Model. It also provides a detailed, lengthy analysis of The Beer Game, which has been used for many years to demonstrate system oscillations when there are feedback loops with delays. The book helps apply systems thinking to real-world issues.

- Ballé, *Managing With Systems Thinking: Making Dynamics Work for You in Business Decision Making*. Ballé's text focuses on applying systems concepts to the workplace. He observes that a typical management reaction to an issue is a myopic, short -term solution instead of a long-term systemic analysis, and that therefore many "solved" problems recur. His basic points include:

1. Detect patterns, not just events.
2. The use of circular causality (feedback loops)
3. Focus on the *relationships* rather than the parts


- Norman, *Systems Thinking: A Product is More Than the Product* (*http://www.jnd.org/dn.mss/systems_thinking_a_.html*). A terrific article applying systems thinking to product development using examples such as the iPod, Kindle, and Mini-Cooper. The article explains that a product is more than just the physical entity; it is the experience of researching, shopping, buying, using, and maintaining the product. For example, the iPod is so successful not only because the physical device is beautiful and functional, but also because the music downloading and listening experiences are pleasurable.


*Self-Organization and Emergence*

- Mano, *Self-Organization in Natural Systems*. Mano presents a variety of examples of self-organization in natural systems, including zebra stripes, leopard spots, sand dune ripples, mud cracks, herding of wildebeests, and honeycomb cell structure. He also explains the forces underlying self-organization.

- Camazine, Deneubourg, Franks, Sneyd, Theraulaz, & Bonabeau, *Self-Organization in Biological Systems*. In this text, the authors describe and explain a variety of self-organized natural structures such as ant trails, the synchronization of fireflies, the schooling of fish, bee honeycomb patterns, and termite cathedrals. They also discuss emergent properties.

- Smolin, *The Self Organization of Space and Time*. This is a mind-expanding article explaining how space and time themselves are self-organizing. Smolin explains how self-organization mechanisms create complexity from simple rules and that imbalance in the fundamental forces (gravity, electromagnetic, strong nuclear, weak nuclear) lead to inhomogeneity and complexity. He argues that the structure of the universe and even its origins are caused by self-organization. The article raises significant questions about the necessity of a prime mover in explaining the structure and existence of the universe.

- Beckenkamp, *The Herd Moves? Emergence and Self-Organization in Collective Actors*. This article focuses on self-organization in the natural sciences. It links self-organization and emergence as well as self-organizational concepts in biology, economics, and sociology. It explains why reductionist thinking does not work for complex systems.

*General Works*

- Midgley (Ed.), *Systems Thinking*. This 2003 4-volume set comprises 76 papers by renowned scholars such as Bertalanffy, Boulding, Wiener, Ashby, Bateson, Forrester, Meadows, Beer, Ackoff, Checkland, Senge, Sterman, and Jackson. However, the set covers a much broader base than just Systems Thinking; it covers most topics associated with systems including ecological modelling, systems theory, cybernetics, applications to society, family therapy, and management. In fact, the editor admits that he included the "broadest possible range of system ideas." Midgley appears to have co-opted the term "Systems Thinking" to include all topics associated with systems, which we think inappropriate. Despite that, there are some excellent papers on reductionism, holism, emergence, self-organization, and complexity. There is no attempt at integration into a common definition or understanding.

- Checkland, *Systems Thinking, Systems Practice*. This book, originally published in 1981, draws a distinction between "hard" systems thinking (for which problems may be formulated by making choices among alternatives to achieve an end) and "soft" systems thinking, such as human activity or social systems, which are poorly structured and often harder to deal with. Checkland includes a history of systems thinking (and of science in general) and notes that science's historical preoccupation with reductionism is an obstacle to systems thinking. He states that "Systems thinking, then, makes conscious use of the particular concept of wholeness captured in the word "system" to order our thoughts," and "Systems thinking implies thinking about the world outside ourselves." He believes that systems thinking is founded on a) emergence and hierarchy and b) communication and control. Checkland presents a 7-step methodology for dealing with real-world soft systems problems and provides examples including the declining performance of a textile firm, mining equipment problems, executing useful and meaningful surveys, and the decision to land a man on the moon before 1970.

- Davidz and Nightingale, *Enabling Systems Thinking to Accelerate the Development of Senior Systems Engineers.* This article helps by providing a definition of systems thinking: Systems thinking involves "Utilizing modal elements to consider the componential, relational, contextual, and, dynamic elements of the system of interest." Its principal focus is on how systems thinking develops in engineers, and it identifies enablers, barriers, and precursors to systems thinking. The authors interviewed 205 senior systems engineers and conclude that the principal mechanisms for developing systems thinking are experiential learning, a supportive environment, and personal characteristics such as personality, curiosity, open-mindedness, and the ability to tolerate uncertainty.

- Maani, *Systems Thinking International.* Maani states that

  "Systems Thinking is a way of thinking about life, work, and the world based on the importance of relationships (interconnections). Systems Thinking also provides a language and a scientific technology for understanding and dealing with complexity and change. Systems Thinking has three aspects. These aspects can be used individually or in combination. They are:

- A way of thinking (paradigm) about the world and relationships. The Systems Thinking Paradigm consists of a set of principles and theories.

- A language for understanding change, uncertainty and complexity. The Systems Thinking language uses diagrams to explain non-linear cause and effect relationships.

- A technology for modeling complex situations underlying business, economics, scientific, and social systems. Systems Thinking modeling tools can be used to create powerful simulation models of organizational situations such as strategy development, process design and re-engineering, and team and organizational learning."

- Maani and Cavana, *Systems Thinking, System Dynamics: Managing Change and Complexity*. This book starts with a fairly conventional definition of "system" and goes on to argue that systems thinking is a paradigm involving the big picture view (including components and their interactions), dynamic thinking, operational thinking including the "physics" of operations, and closed-loop thinking. It states that systems thinking is also a language involving diagrams, a syntax with precise rules, the translation of perceptions into pictures, and an emphais on closed-loop interdependencies. It advocates several tools including causal loop diagrams, stock-and-flow diagrams, computer simulations, learning laboratories, and group model building. Maani and Cavana embrace a 4-tiered Iceberg Model and suggest 5 phases of systems thinking and modeling: 1) Structure the problem, 2) Construct Causal Loop diagrams, 3) Model dynamically, 4) Scenario Planning and Modelling, and 5) Implementation and Organizational Learning using Flight Simulators. The book ends with several case studies including the bird flu pandemic, quality in health services, the New Zealand fishing industry, and telecommunications business strategy.

- Valerdi, *Why Systems Thinking is Not a Natural Act.* Valerdi describes 7 systems thinking competencies:

    1. Ability to define the "universe' appropriately - the system operates in this universe
    2. Ability to define the overall system appropriately - defining the right boundaries
    3. Ability to see relationships - within the system and between the system and universe
    4. Ability to see things holistically - within and across relationships
    5. Ability to understand complexity - how relationships yield uncertain, dynamic, nonlinear states and situations
    6. Ability to communicate across disciplines - to bring multiple perspectives to bear
    7. Ability to take advantage of a broad range of concepts, principles, models, methods and tools - because any one view is inevitably wrong

- Hitchins, *System World* (*www.hitchins.net*), explains systems thinking in terms of 3 generic areas: synthesis, the organismic analogy, and holism:

    "**Synthesis** is the opposite of reduction. Synthesis proposes that the various parts of a complex system cannot exist/survive/operate/ behave/even be considered in mutual isolation. A system comes into existence when the complementary parts are brought together. Each then depends for its very existence on interchanges with the other parts. In turn, this implies that open systems are/have to be active/dynamic. The **organismic analogy** proposes, not that all complex systems are organisms, but rather that, like biological organisms, they behave as unified wholes. Each has a life-cycle, each exhibits growth, stability and death - often sudden, collapsing death. **Holism** proposes that everything within a system is connected/related to -

and affects - everything else, so there is mutual interdependence. Viewing, or even considering, parts on their own is irrational. Systems and their problems have to be viewed as a whole. Holism observes the tendency of the natural world to create 'wholes,' and that a whole may be more than the sum of its parts ..."

Hitchins believes that systems thinking is ".... simply thinking about the world around us, about situations and problems, and how things might/could/should/do work;...thinking about emergent properties, capabilities and behaviours, how they come about, what benefit they might be, what problems they might create… unravelling the inner workings of complex systems… " Hitchins embraces causal loop diagrams and modelling using STELLA and iThink, Interpretive Structural Modelling, $N^2$ Charts, and a rigorous soft systems methodology but he does not embrace the iceberg model.

- Bellinger, *Systems Thinking – A Disciplined Approach* (*http://www.systems-thinking.org/stada/stada.htm*). This article focuses on a suggested approach for developing models to gain an understanding of the underlying structure(s) which give rise to observed patterns of behavior. Bellinger proposes that such an approach consists of the following steps:

  1. Define the Situation
  2. Is Systems Thinking Appropriate?
  3. Develop Patterns of Behavior
  4. Evolve the Underlying Structure
  5. Simulate the Underlying Structure
  6. Identify the Leverage Points
  7. Develop an Alternate Structure
  8. Simulate the Alternate Structure
  9. Develop an Adoption Approach

The author also provides a description of the basic structures and constructs used to model systems in terms of Causal Loop Diagrams (CLDs) and a method for

translating them to Stock and Flow Diagrams at
http://www.systems-thinking.org/stsf/stsf.htm

- Jackson, *Systems Thinking: Creative Holism for Managers.*
This interesting book begins with a conventional definition
of "system," the concepts of holism and reductionism, and
a discussion of hard versus soft systems thinking. It goes
on to outline and critique 10 applied systems approaches:
hard systems thinking, system dynamics, organizational
cybernetics, complexity theory, strategic assumption
surfacing and testing, interactive planning, soft systems
methodology, critical systems heuristics, team syntegrity,
and post-modern systems thinking. The overviews and
critiques are balanced and fair; however, Jackson
concludes with a recommendation for "Total Systems
Intervention" or TSI which seems to argue that no one
approach will address all problems and that one must
therefore pick and choose the combination of approaches
that work best for a situation. This approach is unclear to
us and does not represent an integrated perspective. The
book is excellent, however, in describing and critiquing
several popular systems approaches.

- At the 2015 Conference on Systems Engineering Research,
Arnold and Wade proposed a novel self-referential
description of systems thinking in which they suggest that
systems thinking is itself a system. They then developed a
"Systems Test" for systems thinking definitions: the
definition must describe the purpose, elements, and
interconnections of systems thinking and must identify
systems thinking itself as a *system*. They next compared
systems thinking definitions from 7 different authors and
demonstrated that each definition fails their Systems Test;
however they do identify the following commonalities
among the definitions: interconnections, the understanding of dynamic behavior, systems structure as a cause
of that behavior, and the idea of seeing systems as wholes
rather than parts. Arnold and Wade argue that previous
definitions do not adequately describe what systems

15

thinking does and propose a new definition: "Systems thinking is a set of synergistic analytic skills used to improve the capability of identifying and understanding systems, predicting their behaviors, and devising modifications to them in order to produce desired effects. These skills work together as a system." Although their definition is helpful and their approach is unique, we see no *a priori* reason why definitions of systems thinking must pass their systems test in order to be valid, useful definitions.

- Russell Ackoff's, Herbert Addison's, and Andrew Curley's short book, *Systems Thinking for Curious Managers*, is more about Ackoff's famous f-Laws than about systems thinking, although some consider his f-Laws a distillation of systems thinking concepts. (Some f-Law examples include "The amount of time a committee wastes is directly proportional to its size," "The less sure managers are of their opinions, the more vigorously they defend them," and "Administration, management, and leadership are not the same thing.") Although the f-Laws are sometimes droll and even poignant, they are presented randomly and do not directly advance the understanding of systems thinking. On the other hand, Ackoff's co-authors propose the following working definition of systems thinking: "Systems thinking looks at relationships (rather than unrelated objects), connectedness, process (rather than structure), the whole (rather than just its parts), the patterns (rather than the contents) of a systems, and context. Thinking systemically also requires several shifts in perception, which lead in turn to different ways to teach, and different ways to organize society." The book goes on to discuss feedback loops, tropisms, self-organization, interconnectedness, equifinality, events versus systems, parts versus the whole, the whole in context, mess, analysis versus synthesis, failure to learn, change, aims and intentions, and people.

## What is Systems Thinking? ---Recommendations

There are many different views regarding the definition of Systems Thinking, and as yet there does not seem to be a precise, widely-accepted definition. However, there appear to be common themes that are repeated in many of the sources. This section will attempt to identify and integrate those common themes into a coherent definition.

---

→Systems Thinking is a perspective, a language, and a set of tools.

---

## The Systems Thinking Perspective

Most sources agree that systems thinking is the opposite of linear thinking, and that it focuses on the relationships among system components, as opposed to the components themselves. It is holistic (integrative) thinking instead of analytic (dissective) thinking. The scientific method prevalent in the last 2 centuries has taught us that we must break up complex situations into smaller and smaller pieces to understand them: dissective thinking. While this has great benefits, it also has the great disadvantage of ignoring the relationships among system components; those relationships often dominate systems behavior. Systems thinking requires that we study systems holistically. This holistic thinking involves both spatial and temporal elements, as shown in Figure 2.

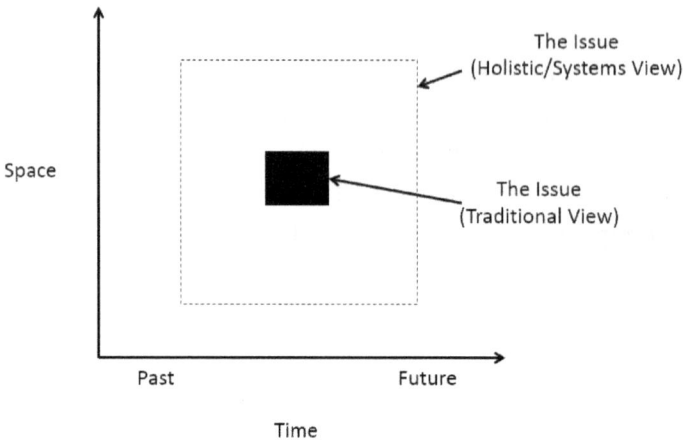

Figure 2. Systems Thinking versus Traditional Views

The space element is often easier to grasp than the time element. But systems thinking requires that we ask: What circumstances and attitudes led to this point? What actions and behavior patterns led to this point? What are the likely attitudes, actions, and patterns going forward? What are the probable reactions of my: allies, enemies, competitors, neutral 3rd parties, and the environment? Systems Thinking thus requires a vision of the future as well as an understanding of the past.

Systems thinking acknowledges that systems are dynamic, and has evolved from the field of General Systems Theory (Bertalanffy). Systems are constantly subject to various forces and feedback mechanisms, some of which are stabilizing and some of which are reinforcing or de-stabilizing. If there are feedback loops with delays, systems may oscillate—examples are one's checking account balance, employee turnover, the national economy, predator-prey populations, or a mass at the end of a spring. This behavior is often counter-intuitive. System dynamics and system dynamics modeling are used to help understand the behavior of systems over time, to identify the driving variables so that system behavior may be positively impacted, and to help predict future states.

It is important to note that systems thinking does not supplant either statistical or reductionist (analytic) thinking; it complements them, as shown in Figure 3:

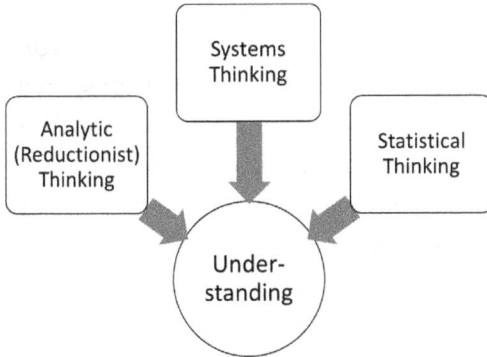

Figure 3. Systems Thinking Complements Analytic and Statistical Thinking

Following Weaver's original explanations, Weinberg (see Figure 1) pointed out that systems thinking deals with **organized complexity** as opposed to **organized simplicity** (which can be dealt with analytically using the laws of physics, for example) and **unorganized complexity** (which can be dealt with statistically using statistical mechanics.) All three approaches provide different but complementary perspectives on gaining more insight into and understanding of the behavior of a system.

Systems Thinking requires that we recognize that in human-designed systems, repeated events or patterns derive from systemic structures which, in turn, derive from mental models. This is clearly depicted in the *Iceberg Model (Figure 4)*, which is a core element of systems thinking:

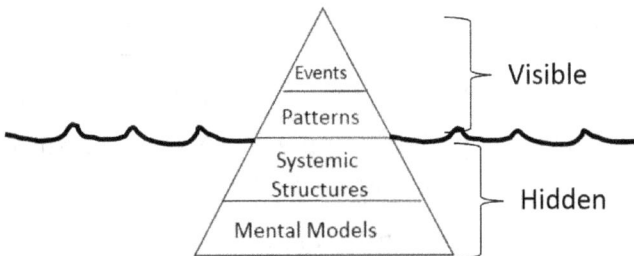

Figure 4. The Iceberg Model

The Iceberg Model argues that events and patterns (which we can observe) are caused by systemic structures and mental models, which are often hidden. Systemic structures are the organizational hierarchy; social hierarchy; interrelationships; rules and procedures; authorities and approval levels; process flows and routes; incentives, compensation, goals, and metrics; attitudes; reactions and the incentives and fears that cause them; corporate culture; feedback loops and delays in the system dynamics; and underlying forces that exist in an organization. Behaviors derive from these structures, which are (in turn) established due to mental models or *paradigms*. A fundamental systems thinking concept is that different people in the same structure will produce similar results—per Deming, the structure causes 85% of all problems; not the people! In order to understand behaviors, we must first identify and then understand the systemic structures and underlying mental models that cause them. (Note: Some versions of the Iceberg Model omit the lowest level, while some add a 5th level at the bottom entitled "container." There are also other versions involving "vision" and "beliefs." We believe that the 4-level model depicted in Figure 4 is the most useful.)

At this point, the Iceberg Model must be modified to distinguish natural systems from human-designed systems (Figure 5):

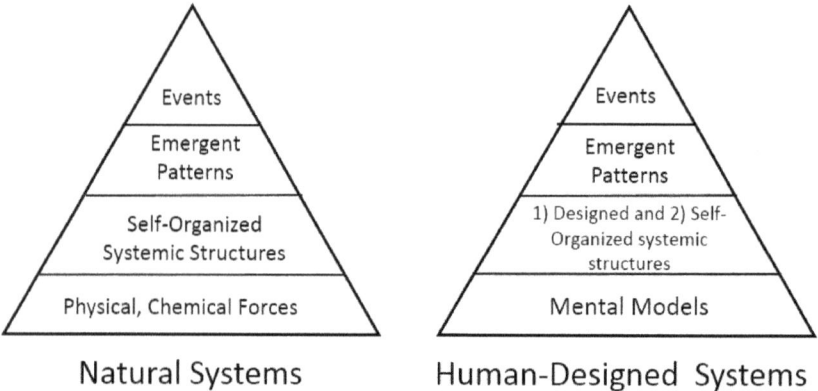

| Natural Systems | Human-Designed Systems |
|---|---|
| Events | Events |
| Emergent Patterns | Emergent Patterns |
| Self-Organized Systemic Structures | 1) Designed and 2) Self-Organized systemic structures |
| Physical, Chemical Forces | Mental Models |

Figure 5. The Iceberg Model Applied to Natural versus Human-Designed Systems

In natural systems, the structures are always self-organized, while in human-designed systems the structures may be either self-organized or designed. But what is self-organization? Camazine explains it well: "Self-Organization is a process in which a pattern at the global level of a system emerges solely from numerous interactions among the lower level components of the system. Moreover, the rules specifying interactions among the system's components are executed using only local information, without reference to the global pattern. In other words, the pattern is an emergent property of the system, rather than a property imposed on the system by an external influence." Thus, self-organization exists if – independent of the intentions or even existence of an organizer or a central plan – regular or arranged patterns emerge from the interactions in the system itself. This concept has significant implications for the origin of life and of the universe itself.

Camazine's definition introduces the concept of *emergence.* Emergent properties are properties of the system as a whole rather than properties that can be derived from the properties of its components. Emergent properties are a consequence of the *relationships* among system components—they can therefore only be assessed and measured once the components have been integrated into a system. This means that one cannot address emergent properties using reductionist thinking. Examples of emergence in natural systems include the flocking of birds, V formations of geese, schooling of fish, ant colony structure, termite "cathedrals", pressure of gases, and entropy or disorder. Examples of emergence in human-designed systems include the meaning of words, traffic jam patterns, reliability, security, usability, countries, and the power of religion to influence behavior. The relationships among system components (and the behaviors and patterns deriving from those relationships) *are additional key elements* of systems thinking.


*Literature Commonalities.* With respect to perspective, then, most system thinking sources agree that systems thinking is the opposite of linear thinking; that it focuses on relationships versus components, and integration versus dissection; that it recognizes and addresses the dynamic nature of systems and that system feedback loops are essential to understanding system dynamic behavior; that systems exhibit self-organization and emergent properties; and that systems thinking has great power in analyzing, understanding, and influencing complex business, socio-economic, and natural problems and behaviors.

*Literature Disparities.* From the systems thinking literature, however, it also seems that there exist two general schools of thought or common themes regarding systems thinking: one school focuses on the Iceberg Model and on the patterns and events that are caused by systemic structures and mental models. This school sees system dynamics as a fundamental element of systems thinking, but does not equate it to systems thinking. The other school focuses on the inter-relationships among system components, the dynamic behaviors that arise therefrom, and system dynamics modeling, and tends to equate systems thinking with system dynamics, but does not embrace the Iceberg Model. We believe that both the Iceberg Model and system dynamics are fundamental to systems thinking. In fact, the causal loops, inter-relationships among components, and dynamic behavior of systems all fall under the "Systemic Structures" level of the Iceberg Model. Those structures are the causative factors behind patterns and events. Thus the Iceberg Model represents a broader context and demonstrates how the underlying structures impact our daily lives in observable ways. It goes beyond dynamics and considers the psychology behind structure.

For example, a systems thinking analyst may attack a complex problem by first constructing a causal loop diagram, and then translating it into a stock-an-flow diagram, and eventually into a dynamic model using iThink or similar software. The model will lead to the identification of key leverage points and ways to impact the system's behavior. But this new knowledge is then useful in affecting the patterns of behavior deriving from the systemic structure, and subsequently to the events that impact people's lives. In addition, the Iceberg Model's attention to mental models will help determine *why* the structures exist and *how* they may be changed.

*The Integrated Model.* Complete systems thinking thus integrates concepts from the Iceberg Model and concepts from causal loop diagrams and dynamic modeling into an overarching framework. This integrated model is depicted in Figure 6.

Events

Emerge — Patterns

Systemic Structures

Mental Models

- Relationships Among Components
- Causal Loop Diagrams
- Stock and Flow Diagrams
- Dynamic Modelling
- Self-Organization

- Culture, Values
- Paradigms
- Root Cause Analysis

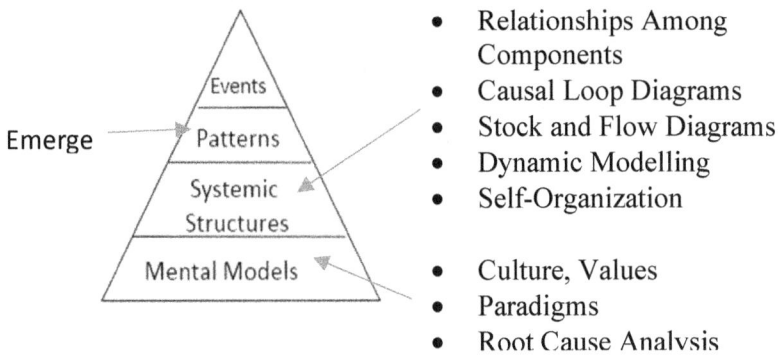

Figure 6. Integrated Model of Systems Thinking

**The Systems Thinking Language.** The Iceberg Model introduces some of the key language of systems thinking: events, patterns, systemic structures, and mental models. Other key words include self-organization, emergence, feedback, system dynamics, and unintended consequences. Causal loop diagrams and stock-and-flow diagrams (described below under "Systems Thinking Tools") are important parts of the systems thinking language and a key means for communicating system components and relationships. A concise summary of systems thinking terms is provided here (some definitions are taken from Kim and are included here with the kind permission of Leverage Networks, Inc. (www.leveragenetworks.com ):

- Accumulator: Anything that builds up or dwindles; for example, water in a bathtub, savings in a bank account, inventory in a warehouse. In modeling software, a stock is often used as a generic symbol for accumulators. An accumulator is also known as a Stock or Level.

- Balancing Process/Loop: Combined with reinforcing loops, balancing processes form the building blocks of dynamic systems. Balancing processes seek equilibrium: They try to bring things to a desired state and keep them there. They also limit and constrain change generated by reinforcing processes. A balancing loop in a causal loop diagram depicts a balancing process.

- Complexity: Characteristic of a system having many components and the multiple ways that those components interact.

- Emergence: Properties of the system as a whole rather than properties that can be derived from the properties of the system components. Emergent properties are a consequence of the *relationships* among system com- ponents. Examples include the flocking behaviour or murmuration of birds, the schooling of fish, the shape of an apple, traffic jam patterns, the concept of countries, and the ability of religion to influence behaviour.

- Events: Things that happen that we can see or observe.

- Feedback: The return of information about the status of a process. For example, annual performance reviews return information to an employee about the quality of his or her work.

- Flow: The amount of change something undergoes during a particular length of time. Examples are the amount of water that flows out of a bathtub each minute, or the amount of interest earned in a savings account each month, which are also called rates.

- Hierarchy: The various levels of organization in a system. In systems, hierarchies often *evolve* from the bottom to the top; stable levels of the hierarchy provide system stability and resilience. Hierarchies also facilitate the evolution of simple systems into complex systems.

- Holism: The theory or philosophy that systems display characteristics that are more than the sum of their parts and that system understanding cannot be attained by analyzing the parts in isolation.

- Leverage Point: An area where small change can yield large improvements in a system.

- Mental Models: paradigms or belief structures that attempt to interpret and/or simplify the universe in which we live. Examples are "An MBA will make you rich," "Incentive compensation increases productivity," and "Girls like Corvettes." Mental models often lead to systemic structures which are either intentional or emergent.

- Patterns: Sets of consistent and recurring *observable* events. Patterns may be physical, behavioral, or mental. Patterns are usually caused by underlying systemic structures and forces.

- Reinforcing Process/Loop: Along with balancing loops, reinforcing loops form the building blocks of dynamic systems. Reinforcing processes compound change in one direction with even more change in that same direction. As such, they generate both growth and collapse. A reinforcing loop in a causal loop diagram depicts a reinforcing process, which is also known as a vicious cycle or a virtuous cycle.

- Self-organization: The tendency of a system to develop structures or patterns without the intervention of a designer or central plan, simply because of the interactions among the system elements. Good examples include the tendency of a free market system to organize into buyers, sellers, traders, and bankers, and the tendency of geese to organize into a V-formation.

- Structural Diagram: Depicts the accumulators and flows in a system, giving an overview of the major structural elements that produce the system's behavior. Structural diagrams are also called flow diagrams or accumulator/flow diagrams.

- Structure: The manner in which a system's elements are organized or interrelated. The structure of an organization, for example, could include not only the organizational chart but also information flows, interpersonal interactions and relationships, rules and procedures, authorities and approval levels, process flows, routes, attitudes, reactions and the incentives and fears that cause them, corporate culture, and feedback loops.

- System: A group of interacting, interrelated, or inter-dependent elements forming a unified whole that attempts to maintain stability through feedback, has boundaries and constraints, and for which the arrangement of the parts is significant. There are both human-designed systems

25

(which serve a specific purpose) and natural systems such as the solar system (which may not have a specific purpose or whose purpose is unknown to us.)

- Systems Thinking: A school of thought that focuses on recognizing the interconnections between the parts of a system and synthesizing them into a unified view of the whole.

- Stock: See Accumulator.

- Unintended Consequences: Results of actions that were nether planned nor foreseen due to a lack of systems thinking. Examples include the negative impact of DDT on the environment, the dramatic increase in organized crime as a result of prohibition, the over-use of antibiotics resulting in antibiotic-resistant bacteria, and the devastation caused by gypsy moths, which were original imported to the United States as a cheaper source of silk.

**Systems Thinking Tools.** There are many systems thinking tools, but not all of them are fundamental or integral to the practice of systems thinking. To identify those that are fundamental, we have established the following criteria:

1. The tool must be widely applicable to most systems, not to a narrow sub-category of systems
2. It must be described in the systems thinking literature
3. The tool must be easy to use and understand without extensive training
4. It must address at least one of the concepts described above under the definition of systems thinking
5. Its principal focus must be on the understanding of *existing* systems as opposed to the design of new systems (which we would describe as a system *design* tool)

We believe that the following eight tools meet these criteria:

- Systems Archetypes
- Behavior Over Time Graphs

- Causal Loops Diagrams with Feedback and Delays
- Systemigrams
- Stock and Flow Diagrams (including Main Chain Infrastructures)
- System Dynamics/Computer Modeling
- Root Cause Analysis
- Interpretive Structural Modeling (ISM)

*Systems Archetypes.* In systems thinking, archetypes are **problem-causing** structures that are repeated in many situations, environments, and organizations. Being facile at identifying them is the first step in changing the destructive structure. There are 10 common archetypes: Accidental Adversaries, Fixes that Fail (policy resistance), Limits to Growth, Shifting the Burden (addiction), The Tragedy of the Commons, Drift to Low Performance (eroding goals), Escalation, The Rich get Richer, Rule Beating, and Seeking the Wrong Goal. These 10 archetypes are very common in business situations, and the literature presents many suggestions for dealing with them. The key is to first identify them.

*Behavior Over Time (BOT) Graphs.* Behavior Over Time graphs plot the values of pertinent system variables over time. They are often useful first steps in developing an understanding of systemic behavior and of how variables inter-relate.

*Causal Loops with Feedback and Delays.* System behavior is usually determined by the presence of reinforcing and balancing processes. These are sometimes obvious (such as the reinforcing process of compound interest) and sometimes not (as in the stabilizing impact of terrorism on international collaboration). In either case, drawing causal loop diagrams helps to see the interrelationships among all system components. These can become quite complicated as cause-and-effect relationships, many of which are hidden (or at least hard to see), are identified. But one of the first steps in attempting to understand system behavior is the construction of a causal loop diagram. Kim and Meadows both present good examples and explanations. An example of a very simple temperature control causal loop diagram is shown in Figure 7:

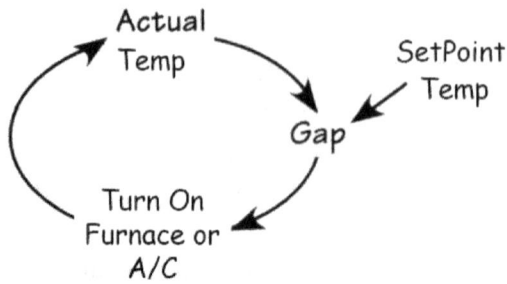

Figure 7. A Simple Causal Loop Diagram

*Stock and Flow Diagrams.* Systems often involve accumulators or stores of "things." The things may be physical quantities such as volume of water, quantity of electric charge, number of rabbits in a field, number of customers of a company, or amount of money in a Certificate of Deposit. They may also be non-physical things such as emotions: love, greed, angst, or lust. In systems, these quantities of things are called stocks. Stocks may increase or decrease due to flows into or out of them. Stock and flow diagrams show the stocks, inflows, and outflows. They are often developed in conjunction with causal loop diagrams, and they are important precursors to system dynamics modeling. Stock and flow diagrams, like causal loop diagrams, are invaluable in understanding system behavior, and Bellinger provides a method for translating causal loop diagrams into stock-and-flow diagrams. In addition, Goodman, Kemeny, and Roberts provide a detailed description of the language of loops and links. A simple stock-and-flow diagram depicting logging impact on a forest (from Meadows) is shown in Figure 8:

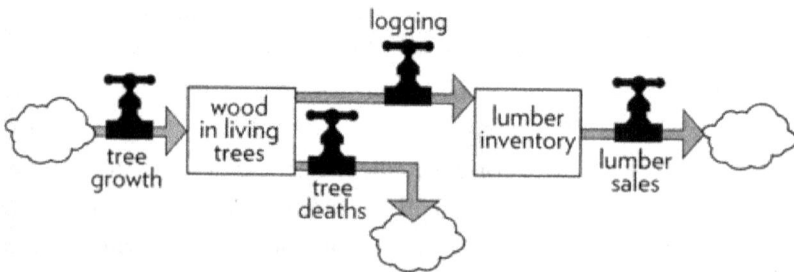

Figure 8. A Simple Stock-and-Flow Diagram (from Meadows)

*Main Chain Infrastructures.* Some stock-and-flow infrastructures are repeated frequently in business and scientific systems. These include human relations, customer, administration, manufacturing, sequential work flow, and queue/server. Primarily used for system dynamics modeling, these main chains are described well in Richmond and provide a head-start for anyone attempting to model system dynamic behavior. An example of a manufacturing main chain infrastructure (from Richmond) is shown in Figure 9:

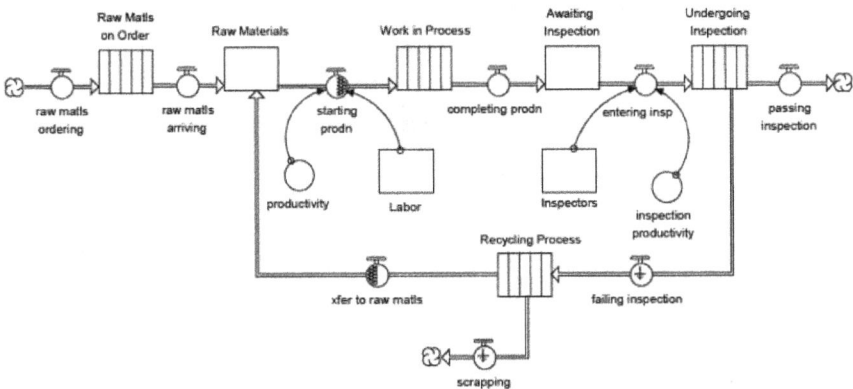

Figure 9. Manufacturing Main Chain Infrastructure (from Richmond)

*Systemigrams.* Derived from the words "Systemic Diagram," systemigrams attempt to translate a system problem (expressed as structured text) into a storyboard-type diagram describing the system's principal concepts, actors, events, patterns, and processes. They typically read from the upper left to the lower right, communicating thereby the chief message of the text. Per Boardman, the diagram is a network comprising nodes, links, flows, inputs, outputs, beginning, and end, and it must fit on a single page (although that page may be quite large.) Figure 10 (from Sauser) shows a beautiful systemigram describing the IED problem as it affects U. S. soldiers. Colors may be used to indicate similar or linked concepts or transformations, or to draw attention to key elements. One can see that although systemigrams contain elements of causal loop diagrams, they are substantially more than that and their main thrust is not feedback loops, but rather telling a story. Although systemigrams are very useful in understanding existing systems, there have been recent attempts (Cloutier et al.) to use them to bridge the gap between

29

systems thinking and MBSE. Details of systemigrams (and more examples) may be found in both Sauser, and Boardman and Sauser.

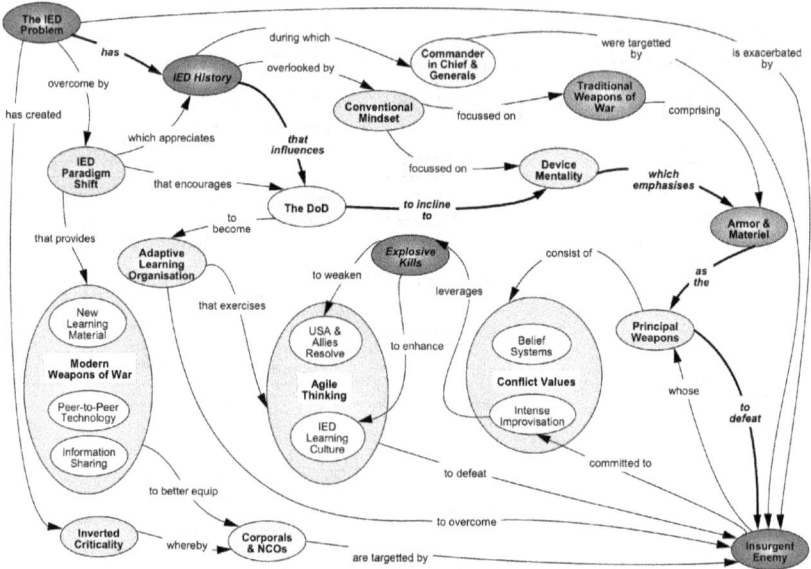

Figure 10. IED Systemigram (from Sauser)

*System Dynamics/Computer Modeling.* System Dynamics is the study and analysis of system behavior over time (feedback loops, time delays, non-linear behavior). System dynamics was originally developed in the late 1950s by Jay W. Forrester of the MIT Sloan School of Management with the establishment of the MIT System Dynamics Group. It is difficult to understand a system without understanding its behavior over time, which is often non-intuitive. Modeling of a system helps understand why the system (company/individual/department) behaves as it does. Modeling also helps identify control points and how one can influence the system. Several software packages are available for systems dynamics modeling, including Stella and iThink from isee Systems, Vensim from Ventana Systems, and Powersim from Powersim AS. A more complete list system dynamics modeling tools can be found at http://en.wikipedia.org/wiki/List_of_system_dynamics_software.

*Interpretive Structural Modeling (ISM)*. ISM is a computer-aided interactive learning process that attempts to identify systemic structures by transforming vague, poorly defined mental models into clear, well-defined graphic representations. ISM begins by first identifying relevant variables and plotting them as points on a graph. Those elements that are related are connected by a directional line. The existence and nature of the relationships are determined by a brainstorming group whose collective judgment determines the final model; it is thus a group learning process. Typical steps in ISM are: 1) Develop a Structural Self-Interaction Matrix (SSIM). In this step, a group of experts plot system elements and determine inter-relationships, indicating them with arrows. 2) In the $2^{nd}$ step a "Reachability Matrix" is developed by using symbols to represent the relationships between elements as unidirectional, bi-directional, or non-existent 3) Step 3 calls for Partition Leveling. First, all elements that are impacted by a particular element (the "reachability set") are identified. Then, all elements that impact that element (the "antecedent set") are identified. An intersection set (representing the intersection of the reachability and antecedent sets) is identified for each element. Those elements for which the intersection set is the same as the reachability set are identified as "Level 1" and are removed from further consideration. Level 1 elements display closed feedback loop impacts; that is every element impacted by the element also impacts the element. The process is repeated until the level of all elements has been determined. 4) A Canonical Matrix is developed by grouping elements of the same level. 5) Based on the Canonical Matrix, a Digraph or hierarchical structure is developed showing the most important factors at the top and less-important factors beneath 6) The ISM Model is developed from the digraph by replacing nodes with verbal descriptions. This technique identifies those elements that are most strongly dependent upon other elements, and also those elements that are the strongest influencers of other elements. The Analytic Hierarchy Process (Saaty, 1980) is often used in conjunction with ISM to assess the relative importance of elements at the same hierarchical level of the ISM Model. The beauty of ISM is that it takes advantage of the knowledge and views of experts and synthesizes them into a system pictogram that clearly identifies the most important elements and relationships in a system. ISM is explained in detail by Warfield (1974), Attri *et al.* (2013), and Lendaris (1980).

*Systemic Root Cause Analysis (RCA).* RCA is a class of problem solving methods aimed at identifying the root causes (not the symptoms) of problems or events. It is especially good for solving problems caused by the system (and many are). Root Cause Analysis is a step by step method that leads to the discovery of a fault's first or root cause, typically starting with the Five "Why's". But according to Mark Galley of *Think Reliability*, "....most organizations stop their RCA too early, at (for example) one of the following: human error, procedure not followed, training less than adequate, or equipment failure." Galley goes on to say that while it is true that these things occur, addressing them is like a Band-Aid. To cure the disease, we must identify and change the *systemic structure* that allowed them to occur by asking specific "Why?" questions. The systems approach to RCA is based on this principle of systemic cause and effect.

Most of us are familiar with the "Oil-On-the-Floor" example. In this scenario a Plant Manager walks into the plant and finds oil on the floor. He calls the Foreman over and asks him why there is oil on the floor. The Foreman indicates that it is due to a leaky gasket in the pipe joint above and that an entire batch of gaskets is defective. The Plant Manager then talks with Purchasing about the gaskets; the Purchasing Manager indicates that they were bought from an unknown vendor because that vendor was the lowest bidder. The Plant Manager then asks the Purchasing Manager why they went with the lowest bidder, and he indicates that was the direction he had received from the VP of Finance. When the Plant Manager asks the VP of Finance why Purchasing had been directed to always take the lowest bidder, the VP of Finance says, "Because you indicated that we had to be as cost conscious as possible!" The Plant Manager is horrified to realize that *he* is the reason there is oil on the plant floor. And in conventional linear thinking, the scenario ends there. **But that's not the solution in systems thinking!** The plant manager could very well conclude that he should be more careful in the future when giving directives, and that he should consider ramifications and how people might react; he might even conclude that he should be more of a systems thinker. But until every manager in the plant understands "unforeseen consequences" and systems thinking, this type of problem will recur!

Systems thinking requires that we address the *system* that allowed the plant manager to give such a directive and that allowed the VP of finance ("be cost-conscious" means "buy from the lowest bidder"-linear!), purchasing manager, etc., to react in such non-constructive, linear-thinking ways. We have found it interesting that very often, it is the *environment* or *corporate culture* that is the systemic root cause of many problems (e.g. the space shuttle Challenger disaster, the Jerry Sandusky child-abuse debacle.)

Several of these tools couple together nicely. Causal Loop Diagrams, for example, are good precursors to stock-and-flow diagrams which, in turn, are helpful in developing dynamic system models using iThink, Stella, or similar software packages. Bellinger (2004) does a nice job explaining how CLDs can be translated into stock-and-flow diagrams. Root cause analysis and archetypes also couple well.

## Conclusions

There is a good deal of literature about Systems Thinking presenting a variety of concepts and viewpoints, many of which are disparate. In this paper we have reviewed some of the key literature and attempted to identify common threads and integrate them into a coherent definition. Systems thinking is 1) a perspective that recognizes systems as collections of components that are all interrelated and necessary, and whose *inter-relationships* are at least as important as the components themselves; 2) a language centered on the Iceberg Model, unintended consequences, causal loops, emergence, and system dynamics, and 3) a collection of tools comprising systemigrams, archetypes, causal loops with feedback and delays, stock and flow diagrams, behavior-over-time graphs, main chain infrastructures, system dynamics/computer modeling, interpretive structural modelling, and systemic root cause analysis.

Systems Thinking provides a great deal of power and value. It can be used to solve complex problems that are not solvable using conventional reductionist (dissective) thinking, because it focuses on the relationships among system components, as well as on the components themselves; those relationships often dominate system performance. It focuses on the properties of the whole that are neither attributable to nor predictable from the properties of the components.

Systems Thinking can be used to explain and understand dynamic non-linear behaviors like the inventory oscillations in supply chain management and the populations of predators and their prey; it can be used to understand complex socio-economic problems, predict behaviors, and identify leverage points (e.g. the instability in Afghanistan and the failure of drinking water systems in Togo); and it can be used to explain and understand the apparently illogical behaviors of individuals, organizations, and even countries (such as the rationale behind John Hinckley's attempted assassination of President Reagan, ISIS's apparently self-destructive behavior, and the failure of Research In Motion to remain competitive in the Smart Phone industry.)

**Future Work**

The real measure of any definition of Systems Thinking is its ability to help understand and address systems issues. Future papers will investigate the application of the above-described description of systems thinking to real-world problems such as root cause analysis of the space shuttle "Challenger" disaster and of the Penn State sex abuse scandal, the demise of Research In Motion, Inc. and of Polaroid, Inc., how to deal with ISIS, the decline of the fin fishing industry off the coast of New England vs the success of Maine's lobster fishing industry, the failure of our domestic drug policies, the British Petroleum Gulf of Mexico oil spill, the 2008 U.S. economic bailout, and the success of Emperor Palpatine of *Star Wars*.

*Acknowledgement*

*We would like to thank the following for their suggestions regarding key systems thinking literature and an optimal approach in analyzing the references: Gene Bellinger, Bob Cavana, Heidi Davidz, Jamshid Gharajedaghi, Michael Goodman, Derek Hitchins, Harold "Bud" Lawson, Donna Rhodes, and Jerry Weinberg.*

# REFERENCES

Ackoff, Russell L., with Herbert J. Addison and Andrew Carey, *Systems Thinking for Curious Managers*, Triarchy Press Limited, Axminster, UK, 2010.

Anderson, Virginia, and Johnson, Laura, *Systems Thinking Basics From Concepts to Causal Loops*, Pegasus Communications, Inc., Cambridge, 1997.

Arnold, Ross D., and Wade, Jon P., "A Definition of Systems Thinking: A Systems Approach," 2015 Conference on Systems Engineering Research, *Procedia Computer Science* **44**, 2015, 669-678.

Aronson, Daniel, "An Overview of Systems Thinking," http://resources21.org/cl/files/project264_5674/OverviewSTarticle.pdf :

Attri, R., Dev, N., and Sharma, V., "Interpretive Structural Modelling (ISM) Approach: An Overview," *Research Journal of Management Sciences* **2**, 2013, 3-8.

Ballé, Michael, *Managing With Systems Thinking*, McGraw-Hill, Columbus, OH, 1996.

Beckenkamp, Martin, "The Herd Moves? Emergence and Self-Organization in Collective Actors," Max Planck Institute for Research on Collective Goods, Bonn, 2006 http://www.coll.mpg.de.

Bellinger, Gene, "Systems Thinking – A Disciplined Approach," http://www.systems-thinking.org/stada/stada.htm.

Bellinger, Gene, "Translating Systems Thinking Diagrams to Stock & Flow Diagrams," 2004, http://www.systems-thinking.org/stsf/stsf.htm.

Bertalanffy, "An Outline for General Systems Theory," *British Journal for the Philosophy of Science* **1** (2), 1950.

Bertalanffy, Ludwig von, *General System Theory*, George Braziller, NY, 1969 (revised 1976).

Blanchard, Benjamin and Fabrycky, Wolter, *Systems Engineering and Analysis*, 5th Ed., Prentice Hall, Upper Saddle River, NJ, 2011.

Boardman, John and Sauser, Brian, *Systems Thinking: Coping With 21st Century Problems*, CRC Press, Boca Raton, 2008.

Camazine, Scott, Jean-Louis Deneubourg, Nigel R. Franks, James Sneyd, Guy Theraulaz, & Eric Bonabeau, *Self-Organization in Biological Systems*, Princeton University Press, 2001.

Checkland, Peter, *Systems Thinking, Systems Practice,* John Wiley and Sons, New York, 1981.

Cloutier, Robert, Brian Sauser, Mary Bone, and Andrew Taylor, "Transitioning Systems Thinking to Model-Based Systems Engineering: Systemigrams to SysML Models," *IEEE Transactions on Systems, Man, and Cybernetics: Systems,* **45** (4), 2015.

Davidz, H.L. and Nightingale, D.J., "Enabling Systems Thinking to Accelerate the Development of Senior Systems Engineers," *Systems Engineering* Vol 11 No. 1, 2008.

Deming, W. Edwards, *Out of the Crisis*, The MIT Press, Cambridge, MA, 1982.

Galley, Mark, "Think Reliability: Investigation Basics—The Systems Approach," Houston, 2014 http://www.investigationbasics.com/Systems-Approach.aspx .

Gharajedaghi, Jamshid, *Systems Thinking—Managing Chaos and Complexity: A Platform for Designing Business Architecture,* 2nd Ed., Elsevier, Burlington MA, 2006.

Goodman, Kemeny and Roberts, "The Language of Systems Thinking: 'Links' and 'Loops'," The Society of Organizational Learning, https://www.solonline.org/?page=Tool_LinksLoops&hhSearchTerms=%22systems+and+thinking%22.

Haines, Stephen, *The Systems Thinking Approach*, St. Lucie Press, Boca Raton, 2000.

Hitchins, Derek, *Systems World*, www.hitchins.net.

Jackson, Michael C., *Systems Thinking: Creative Holism for Managers*, John Wiley and Sons, Ltd., Chichester, 2003.

Kauffman, Draper L. Jr., *Systems One: An Introduction to Systems Thinking*, Future Systems Inc., S. A. Carlton, Minneapolis, 1980.

Kim, Daniel H., *Introduction to Systems Thinking*, Pegasus Communications (now Leverage Networks Inc., www.leveragenetworks.com), 1999, ISBN 1-883823-34-X.

Kossiakoff, Alexander, Sweet, William, Seymour, Samuel, and Biemer, Steven, *Systems Engineering: Principles and Practice*, 2nd Ed., Wiley, Hoboken, 2011.

Lawson, Harold, *A Journey Through the Systems Landscape*, College Publications, London, 2010.

Lendaris, George G., "Structural Modeling—A Tutorial Guide," *ISEE Transactions on Systems, Man, and Cybernetics*, **SMC 10** (12), December 1980.

Leverage Networks, Inc., www.leveragenetworks.com. Resources, services, and networking covering systems thinking and system dynamics.

Maani, Kambiz, (New Zealand), Systems Thinking International, http://www.sys-think.com/.

Maani, K. E., & Cavana, R. Y., *Systems Thinking, System Dynamic: Managing Change and Complexity* (2nd Ed.), Pearson: Prentice Hall, 2007.

Mano, Jean-Pierre, "Self-Organization in Natural Systems," Institut de Recherche de l'informatique de Toulouse, TFG-AOSE 01.07.04 Roma, 2004. PowerPoint, accessed 7/13/2011.

Meadows, Donella H., *Thinking in Systems: A Primer*, Chelsea Green Publishing, White River Junction, VT, 2008.

Norman, Don, "Systems Thinking: A Product is More Than the Product," http://www.jnd.org/dn.mss/systems_thinking_a_.html.

Pyster, A., D. Olwell, N. Hutchison, S. Enck, J. Anthony, D. Henry, and A. Squires (eds). 2012. *Guide to the Systems Engineering Body of Knowledge (SEBoK) version1.0.* Hoboken, NJ: The Trustees of the

Stevens Institute of Technology ©2012. Available at: http://ww.sebokwiki.org.

Richmond, Barry, *An Introduction to Systems Thinking with iThink*, isee Systems, 2004.

Saaty, Thomas, *The Analytic Hierarchy Process*, McGraw Hill, New York, 1980.

Sauser, B., *Building Systemic Diagrams; Systemigrams*, Stevens Institute, https://www.stevens.edu/sse/sites/default/files/Systemigram_Overvi ew.pdf, accessed 9 March 2015.

Senge, Peter, Kleiner, Art, Roberts, Charlotte, Ross, Richard, and Smith, Bryan, *The Fifth Discipline Fieldbook,*. Doubleday, New York, 1994.

Senge, Peter, *The Fifth Discipline*, Doubleday, New York, 1990; revised 2006.

Smolin, Lee, "The Self Organization of Space and Time," *Phil. Trans. R. Soc. Lond.* A, 2003, 1081-1088.

Sterman, John, The Beer Game, "Teaching Takes Off: Flight Simulators for Management Education. 'The Beer Game,'" http://web.mit.edu/ jsterman/www/SDG/beergame.html.

Sweeney, Linda, and Meadows, Dennis, *The Systems Thinking Playbook*, Chelsea Green Publishing, White River Junction, VT, 1995.

Valerdi, Ricardo, "Why Systems Thinking is Not a Natural Act," MIT SDM Systems Thinking Webinar Series, August 22, 2011.

Warfield, John N., *Structuring Complex Systems (Monograph No. 4)*, Battelle Memorial Institute, Columbus OH, 1974.

Weaver, W., "Science and Complexity," *Am. Scientist 36*, 1948, 536-544.

Weinberg, Gerald, *An Introduction to General Systems Thinking*, Dorset House Publishing, New York, 1975, 2001.

Wikipedia, "List of Systems Dynamics Software," http://en.wikipedia. org/wiki/List_of_system_dynamics_software.

# The Meaning of "Structure" in Systems Thinking

Jamie P. Monat and Thomas F. Gannon

*Reprinted from Systems*, **Vol. 11, 92, 2023,**
https://doi.org/10.3390/systems11020092

**Abstract:** "Systemic structure" is an oft-used term in Systems Thinking. However, different authors use different, sometimes conflicting definitions of "systemic structure," many of which are nebulous, and therefore its meaning is not clear. In this paper, we review the various definitions and interpretations and develop a logical, practical definition that may be applied to develop a deep understanding of system behavior: in Systems Thinking, "structure" is the cause-and-effect manner in which system components interrelate to yield system behavior; and the rules, laws, protocols, procedures, policies, and incentives/rewards that govern those interactions.

**Keywords:** Systems Thinking; structure

## 1. Introduction and Background

There are several different versions of Systems Thinking, many of which use the term "structure." This work focuses on those versions that embrace either the Iceberg Model, Causal Loop Diagrams, system dynamics, or the existence of a link between underlying forces (such as mental models) and patterns to help explain the behavior of simple, complicated, and complex systems. In Systems Thinking, the Iceberg Model (details available in [1]) posits that systemic structure lies between underlying forces (such as Mental Models) and patterns, as depicted in Figure 1.

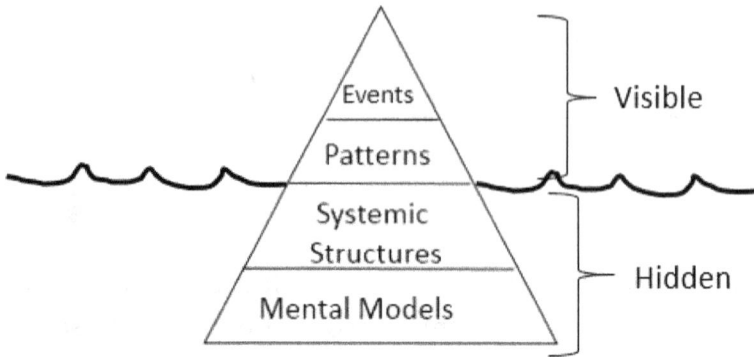

**Figure 1.** The Iceberg Model.

The model argues that in human-designed systems, structures form as a result of Mental Models and that patterns, in turn, form as a result of structure. In natural systems, underlying forces such as gravity, electromagnetism, centrifugal force, and hydrophilicity replace "Mental Models" as the lowest level of the iceberg.

Conventional definitions of "structure" include:

1. the arrangement of and relations between the parts or elements of something complex [2].

2. the arrangement of particle or parts in a substance or body; arrangement or interrelation of parts as dominated by the general character of the whole [3]

3. the way in which the parts of a system or object are arranged or organized, or a system arranged in this way [4]

4. the mode of building, construction, or organization; arrangement of parts, elements, or constituents [5]

5. The structure of something is the way in which it is made, built, or organized [6].

And if one researches synonyms for "structure," Google responds with "configuration," "arrangement," or "organization."

But these definitions are not clear, nor do they provide actionable descriptions. For example, the meanings of "organization," "arrangement," "mode," and "way" are not clear: is it the physical arrangement of system components? Or is it the sequential organization? Or perhaps the reporting relationships in an organization? A better, Systems Thinking-specific definition is required.

## 2. Literature Review

Several researchers have proposed definitions of system structure. We are not aware of any empirical studies that have been done; most of the following definitions represent the opinions and conceptual analyses of the authors.

Daniel Kim [7] states that "Systemic structures are the ways in which the parts of a system are organized. These structures actually generate the patterns and events we observe. Structures can be physical (such as the way a workspace is organized, or the way a machine is built) as well as intangible (such as the ways employees are rewarded, or the way shift changes are timed.)" In this definition, the meaning of the word "way" Is not clear. Does "way" refer to the physical relationships among components, the sequential relationships, the reporting relationships, the authority or power relationships, or something else?

Senge et al. [8] say that "structure is the pattern of interrelationships among key components of the system:

1. organizational hierarchy

2. process flows

3. attitudes

4. perceptions

5. product quality

6. the ways decisions are made"

Their definition refers to structure as a *pattern*. In addition, most Systems Thinking experts would categorize "attitudes" and "perceptions" as mental models instead of structure. And can "product quality" be a system component?

Meadows [9] states that "structure is the system's interlocking stocks, flows, and feedback loops. System structure (feedback loops and underlying forces) is the source of system behavior. System behavior reveals itself as a series of events over time (a pattern!)." This definition includes not only the feedback loops that interrelate system components, but also some system components themselves as well as underlying forces, which we believe are not part of structure but instead are causative factors that *yield* structure.

Monat and Gannon [1] state that structure is "the manner in which a system's elements are organized or interrelated. The structure of an organization, for example, could include not only the organizational chart but also information flows, interpersonal interactions and relationships, rules and procedures, authorities and approval levels, process flows, routes, attitudes, reactions and the incentives and fears that cause them, corporate culture, and feedback loops." Here again, the meaning of the word "manner" is not clear. In addition, Monat and Gannon include some mental models and behaviors in their definition of structure.

Stillwell [10] argues, "when a pattern of transactions occurs over a period of time, it creates a structure that becomes the "cultural norm"—a climate of trust or mistrust. In a reinforcing process, our behaviors strengthen the cultural norm, which strengthens the behaviors, and so on." This definition argues that patterns cause structure when, in fact, the iceberg model suggests that it's the other way around: structure causes patterns.

Spirkin [11] states, "......When studying the content of an object, we enumerate its elements such as, for example, the parts of a certain organism. But we do not stop at that, we try to understand how these parts are coordinated and what is made up as a result, thus arriving at the structure of the object. Structure is the type of connection between the elements of a whole. It has its own internal dialectic. Wholeness must be composed in a certain way, its parts are always related to the whole. It is not simply a whole but a whole with internal divisions. Structure is a composite whole, or an internally organised content. ........ Structure is an extremely abstract and formal concept. Structure

42

implies not only the position of its elements in space but also their movement in time, their sequence and rhythm, the law of mutation of a process. So structure is actually the law or set of laws that determine a system's composition and functioning, its properties and stability." This definition alludes to the organizational relationships among system components and also the rules that govern the system's behavior. But the meanings of "how" and "type" are not clear.

Karash [12] states that "structure is the network of relationships that creates behavior. The essence of structure is not in the things themselves but in the relationships of things. By its very nature, structure is difficult to see. As opposed to events and patterns, which are usually more observable, much of what we think of as structure is often hidden. We can witness traffic accidents, for example, but it's harder to observe the underlying structure that causes them." This definition argues that structure causes events and patterns and that structure is the nature of system component relationships. The meaning of "nature" Is not clear.

Gharajedaghi [13] states that "Structure defines components and their relationships." This tells what structure does but does not elucidate what stricture is.

Mcnaughton [14] states that "The system structure or pattern of organization represents a logical model of the systems for the system-of-interest. This logical model is independent of any specific physical realization of any of the systems. This logical model may also be called a conceptual model of the system-of-interest." This definition equates systemic structure to both a pattern of organization and a logical model.

Austin [15] says, "The structure of a system contains:

> 1. Components. Components are the operating parts of a system consisting of input, process, and output. Each system component may assume a variety of values to describe a system state.

> 2. Attributes. Attributes are the properties of the components in the system.

3. Relationships. Relationships are the links between the components and attributes."

This definition includes system components and their attributes as well as the component relationships.

Barile and Saviano [16] provide the following definitions:

1. "Structure: A set in which the elements are qualified as components recognized as having the capacity to contribute to perform specific functions (necessary to carrying out specific roles in the context of an emerging system). The components can be put in relation respecting specific constraints (rules).

2. Actual structure: Set of physical, concrete components, with a known function provided with a connecting mechanism or linker device predisposed for linking up other components."

This definition purports that systemic structure is the physical system components and the mechanisms that interconnect them; however the nature of the connecting mechanism is not specified.

Anderson and Johnson [17] ask, "....what is structure, exactly? The concept is difficult to describe. In simplest terms, structure is the overall way in which the system components are interrelated—the organization of the system. Because structure is defined by the *interrelationships* of a system's parts, and not the parts themselves, structure is invisible." They further state that "Thinking at the structural level means thinking in terms of causal connections. *It is the structural level that holds the key to lasting, high-leverage change.*" This definition notes that structure involves cause-and-effect relationships among system components.

Senge [18] *says,* "....the structural explanation.....focuses on answering the question, "What causes the patterns of behavior?" In the beer game, a structural explanation must show how orders placed, shipments, and inventory interact to generate the observed patterns of instability and amplification........structure produces behavior.... ...... structure in human systems includes the "operating policies" of the decision makers in the system.....structures are made up of beliefs and assumptions, established practices, skills and capabilities, networks of relationships, and awareness and sensibilities----in other words, the elements of the deep learning cycle." This definition notes that structure causes behavior patterns and that it comprises an array of concepts from beliefs to skills to practices to sensibilities. Later, Senge says, "In human systems, structure includes how people make decisions—the "operating policies" whereby we translate perceptions, goals, rules, and norms into actions."

Stroh [19] says that "...systems structure includes tangible elements such as pressures, policies, and power dynamics that shape performance. It also includes intangible forces such as perceptions (what people believe or assume to be true about the system) and purpose (the actual versus espoused intentions that drive people's behavior)."

Cabrera and Cabrera [20] describe "structure" in a variety of ways ranging from "patterns" to "simple rules that a thing follows" to "grammar and syntax" to "hidden contextual structure that contributes to meaning" to the physical or geometric relationships among system components.

Cabrera et al. [21] argue that "structure" is

- Action-reaction relationships
- Feedback loops
- Identity-other distinctions
- Part-whole systems
- Point-view perspectives

Their empirical validation of the Distinctions-Systems-Relationship-Perspectives (DSRP) model of Systems Thinking is useful but does not fully clarify the meaning of "Structure."

*Commonalities*: Many of these "definitions" assume that the reader already has some concept of systemic structure and define it only obliquely. Most of them explain "structure" in terms of the relationship among system components. *Discrepancies*: However, some authors define "structure" as attitudes and perceptions while other define it as interlocking stocks, flows, and feedback loops while still others define it as system component spatial position, motion, sequence, and mutation laws. Some resolution is required. A plot of the most common descriptors is presented in Figure 2.

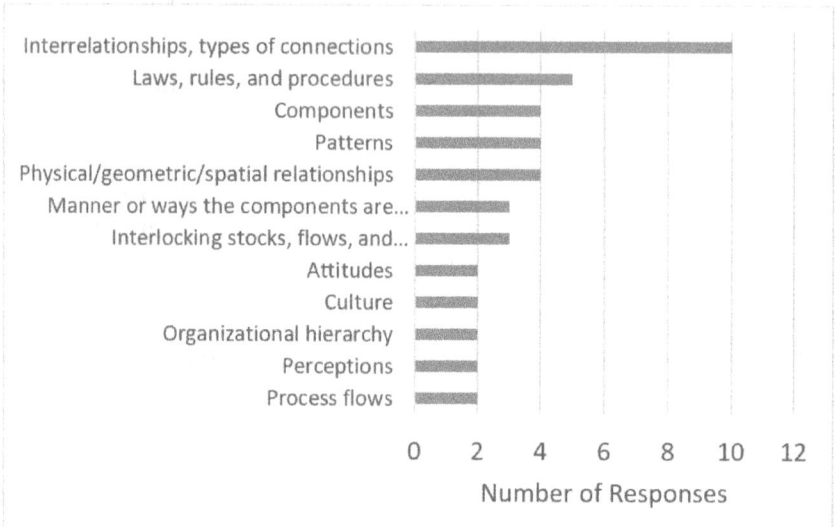

Figure 2. The most common literature descriptors of "Structure." (Only those that appeared two or more times are shown.)


## 3.Discussion

### 3.1 Theoretical Underpinnings

With so many disparate and nebulous definitions of systemic structure, it would be useful to develop a crisp definition that can be easily applied to the analysis and understanding of systems. We may start this development by identifying several widely-accepted Systems Thinking tenets.

Tenet 1: "Structure" must link *underlying forces* (either mental models in human-designed system or natural forces such as gravity and electromagnetism in natural systems) to *patterns* as shown in the Iceberg Model (Figure 1)—that is, structure must explain how underlying forces eventually result in systemic patterns [1, 7, 8, 17, 22, 23].

Tenet 2: "Structure" involves a description of "the way" that or "the manner" in which system components interrelate [1, 7, 11, 16, 17]. The meanings of "the way" and "the manner," however, are not clear, as some references mention physical or geometric relationships while other references cite sequential, temporal, or organizational relationships. A crisp definition of "structure" must clarify what is meant by "the way" that system components interrelate.

Tenet 3: Systemic "structure" must be consistent with and explain the shape of the system's Behavior-Over-Time (BOT) plots [1, 8, 23-28]. Indeed, it is frequently explicitly stated or inferred [7, 8, 12, 17, 28] that a system's BOT plots *reveal* its structure.

Therefore, bearing these fundamental tenets in mind, "structure" must describe how system components interrelate, that is, the impact of each component on the other system components, in a way that explains how and why the systemic patterns arise. Clearly, the description of component interrelationships must be more than just spatial, hierarchical, temporal, sequential, or reporting, because those relationships do not fully explain the impact of one system component on another. To explain the impact of one component on another, one must understand the *cause-and-effect relationships* among the components.

A proposed definition that satisfies the 3 fundamental tenets described above is: In Systems Thinking, "structure" is the cause-and-effect manner in which the system components interrelate to yield system behavior; and the rules, laws, protocols, procedures, policies, and incentives/rewards that govern those interactions. Exactly which components impact other components in cause-and-effect relationships may be depicted in either causal loop or stock-and-flow diagrams. But to completely describe structure, we must show not only the component cause-and effect interrelationships in a Causal Loop Diagram (CLD) or Stock-and-Flow Diagram (S & F,) but also explain how and why those relationships exist, viz. the rules of interaction.

For example, for the case of a thermostat controlling room temperature, the causal loop diagram shown in Figure 3 is appropriate.

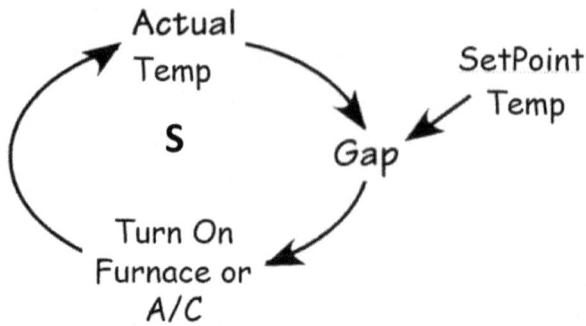

**Figure 3.** Furnace-Thermostat Causal Loop Diagram.

The CLD represents the first half of the description of the system's structure. To complete the description, we add the following *rules of interaction*: "The temperature Gap in the above CLD will cause a thermostat (which detects the gap) to turn the furnace on if the Actual Temperature is below the Setpoint Temperature; or it will cause the A/C to come on if the Actual Temperature is above the Setpoint Temperature. This causes the Actual Temperature to come closer to the Setpoint Temperature, reducing the gap and stabilizing the system." This description elucidates the *cause-and-effect relationships* among the system components and clearly relates the underlying forces (heat transfer and control via a boiler or an air conditioner, and a thermostat to control room temperature) to the observed pattern (a behavior-over-time plot would show the actual temperature smoothly asymptoting to the set-point temperature).

A full description of systemic structure must include not only a CLD or Stock-and-Flow diagram, but also the rules, laws, protocols, procedures, policies, and incentives/rewards that govern the component interactions, because two different systems may have identical CLDs but be governed by different rules. For example, Figures 4 and 5 below show a CLD and Stock-and-Flow diagram, respectively, for population growth. These same diagrams are also accurate for growth of a savings account, spread of a disease, spread of plant seeds, and the growth of ice thickness on a pond. Although the *rules of interaction* for the first 3 examples are all similar (stock growth rate = stock quantity x some efficiency factor) the rules of interaction for ice growth are different: stock growth rate = efficiency factor/(stock value.) The Behavior-Over-Time plots of the first 3 examples show exponential growth (Figure 6) while that for ice thickness (Figure 7) displays a square-root dependence. Thus a full description of systemic structure must include the rules, laws, protocols, procedures, policies, and incentives/rewards that govern the component interactions, as well as either a CLD or Stock-and-Flow diagram.

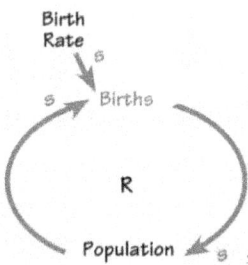

Figure 4. CLD for Population Growth, Savings Growth, Disease Spread, Seed Dispersal, and Ice Thickness Growth

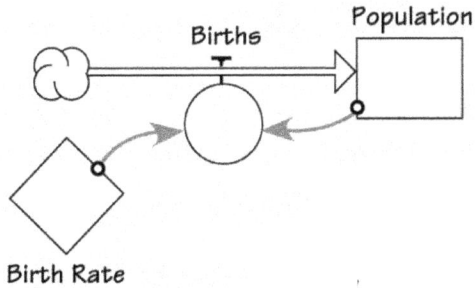

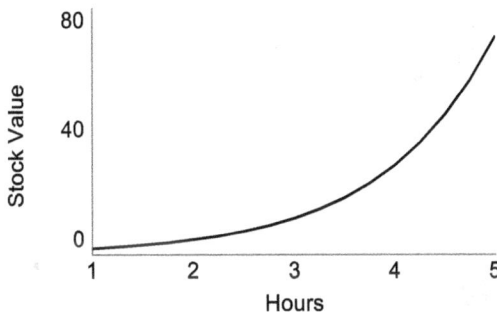

Figure 5. Stock-and-Flow Diagram for Population Growth, Savings Growth, Disease Spread, Seed Dispersal, and Ice Thickness Growth

Figure 6. Exponential Growth

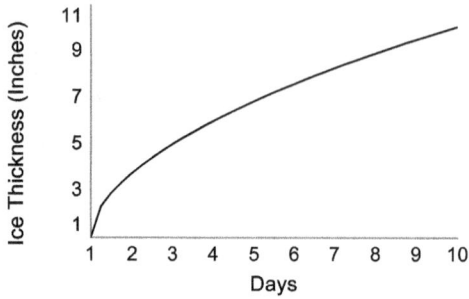

Figure 7. Square Root Growth

Systems Thinking structure is **not** the architecture showing how a building, bridge, or spaceship is built as in Figure 8 (which shows the *geometric* relationships of the system components); the way that a poem or piece of music is configured as shown in Figure 9 (which shows the *sequential* relationships of the components); the *geometric* relationships of the planets in the solar system as shown in Figure 10; or the organizational structure of a corporation as shown in Figure 11 (which indicates *reporting* relationships). Although these conventional structures provide useful information, they do not show the *cause-and-effect relationships* among the system components.

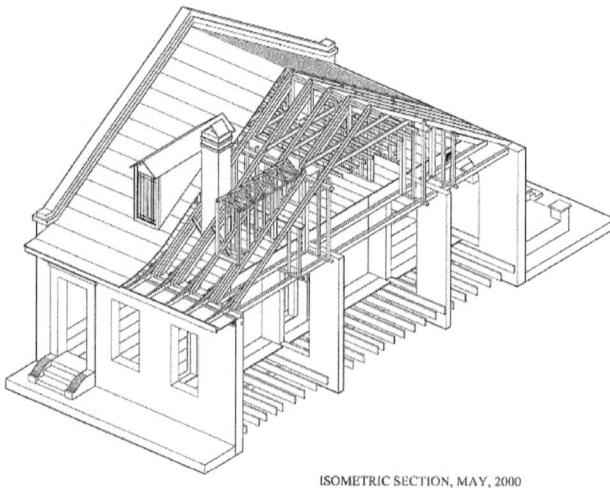

ISOMETRIC SECTION, MAY, 2000

**Figure 8.** House Structure Showing Geometric Relationships of Components [29].

50

**Figure 9.** Song Structure Showing Sequential Relationships of Components.

**Figure 10.** Solar System Structure Showing Geometric Relationship of Components "The Solar System PIA 10231, mod 02, is licensed by Flickr under Creative Commons attribution 4.0 International CC by 4.0" [30].

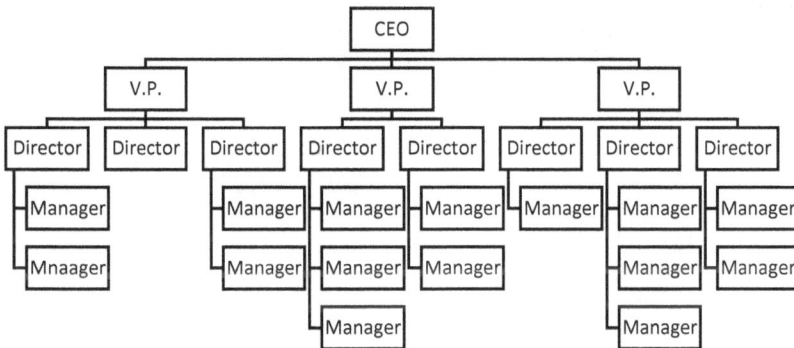

**Figure 11.** Corporate Structure Showing Authority/Reporting Relationships of Components.

Several tools and techniques are available to help determine a system's structure. One approach is to draw a CLD and ask, "What is the cause and effect of one system component on another? What rules govern their relationships?" Another approach requires examination of the Iceberg Model for the system and asking what structure would link underlying forces/mental models to patterns of system behavior. One of the best ways to determine structure is to look at a BOT plot (which shows patterns) and ask, "What would cause that behavior?" It is well-known [7, 8] for example, that exponentially increasing BOT plots indicate reinforcing feedback loops; Behavior-Over-Time plots that are steady or converge indicate stabilizing, balancing, or negative feedback loops; BOT plots that neither converge nor diverge indicate ineffective or absent feedback loops; and that BOT plots that oscillate ˙ ˙˙cate feedback loops with delays (Figures 12-15.)

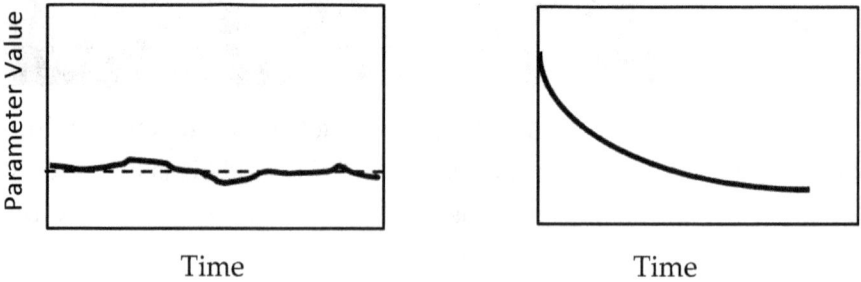

**Figure 12.** Behavior-Over-Time Plots that are steady or converge indicate stabilizing, balancing, or negative feedback loops.

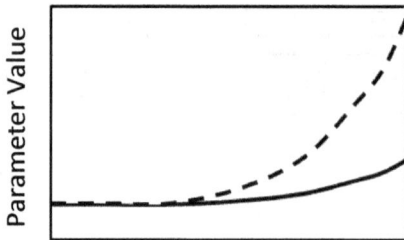

Time**Figure 13.** Behavior-Over-Time Plots that diverge over time indicate reinforcing or positive feedback loops.

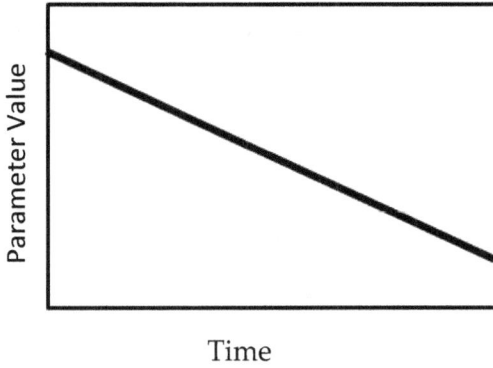

**Figure 14.** Behavior-Over-Time Plots that neither converge nor diverge indicate ineffective or absent feedback loops.

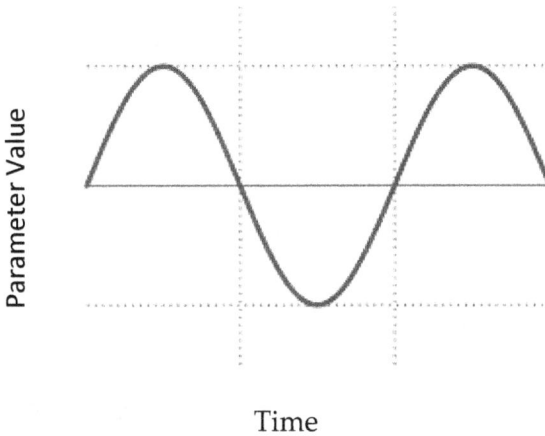

**Figure 15.** Behavior-Over-Time Plots that oscillate indicate feedback loops with delays.

Senge et al. [8] provide several additional common BOT plots that may be used to infer structure. Thus the system's BOT plot provides excellent clues regarding the systemic structure.

### 3.3 Examples

The Iceberg Model is a convenient construct within which to provide examples of systemic structure; it shows how structure is caused by underlying forces and how structure yields patterns. In this section we provide examples of structure in natural systems, human-designed systems, and business systems.

*Example 1.  Structure in Natural Systems: The Spiral Pattern of Scales on a Pine Cone*

*The Pattern*: Spirals of scales in 2 directions on pine cones (Figure 16)

**Figure 16.** Spiral Scale Patterns on Pine Cone. The oscillation suggests a feedback loop with delays.

*Underlying Forces*: The physical/chemical responses to growth hormone concentration

The Structure:

a) The Causal Loop Diagram (CLD; Figure 17):

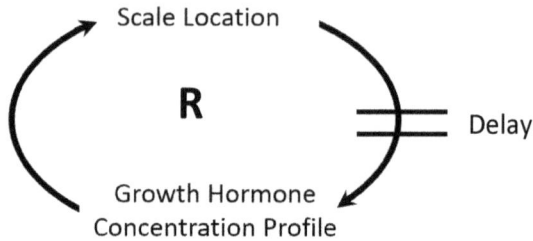

*Figure 17.* Pine Cone CLD Showing the Structure. The scale location impacts the local concentration of growth hormone; the local concentration of growth hormone impacts the scale location in a reinforcing feedback loop.

*b) The Rules of Interaction*: **New buds form around the meristem where there is the highest concentration of growth hormone.** But when a new bud forms, the growth hormone is depleted at that site so that the next bud will form far from the previous bud. The radial growth outward from the meristem along with the growth hormone concentration profile result in a new flake forming at a fixed angle from each previous flake. The net result is a spiral.

*Example 2. **Structure in Human-Designed Systems: Individual Weight Control***

*The Pattern:* An individual's weight oscillates around some fixed value (Figure 18).

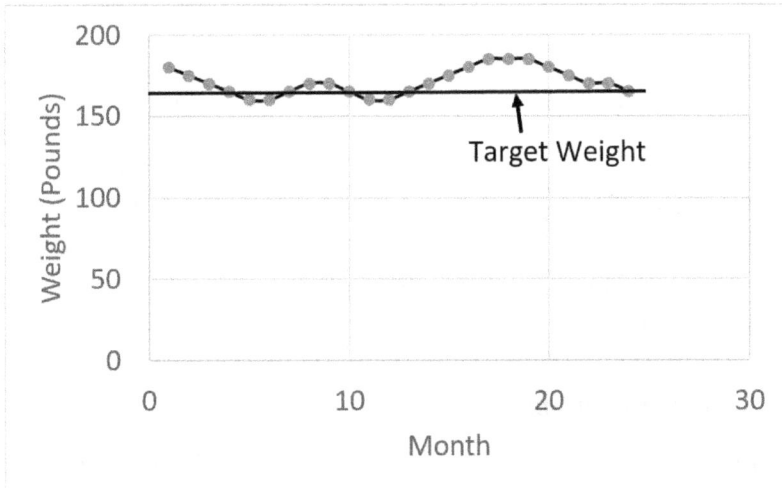

**Figure 18.** Behavior Over Time Graph Showing an Individual's Weight Oscillation around a Target Weight. The curve suggests a stabilizing feedback loop with delays.

*Underlying Forces:* Belief that there is an ideal weight with respect to health, appearance, and well-being (a mental model) and conservation of energy/mass principles (physical/chemical laws) relating to caloric consumption, eating, and exercise

The Structure:

    a) The Causal Loop Diagram (CLD; Figure 19):

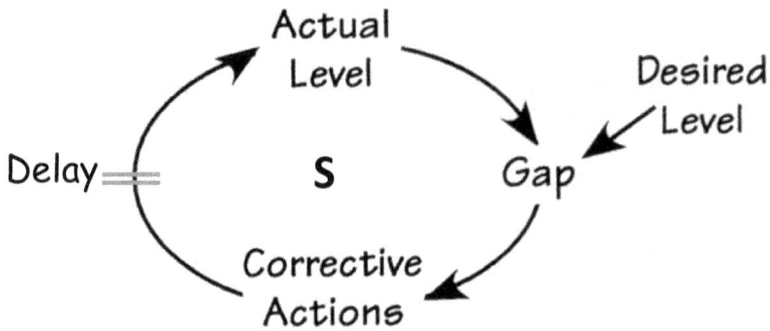

*Figure 19.* Individual's Weight Control CLD Showing the Structure.

*b) The Rules of Interaction*: When an individual's actual weight differs from his desired ideal weight, he will adjust his food consumption and exercise level to bring his weight closer to ideal. The delay between corrective actions and results yields an oscillation.

### Example 3.  Structure in Business Systems:

*The Pattern:* Exponential growth in sales of a capacity-limited product as shown in the BOT plot of Figure 20.

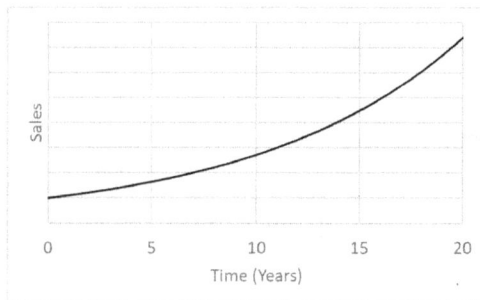

**Figure 20.** Exponential Sales Growth Due to reinvestment of Profits. The exponential growth suggests a reinforcing feedback loop.

*Underlying Forces (Mental Models):* If product demand is so high that sales are limited by production capacity, then increasing capacity will increase sales. But it costs money to increase production.

The Structure:

a) The Causal Loop Diagram (CLD; Figure 21):

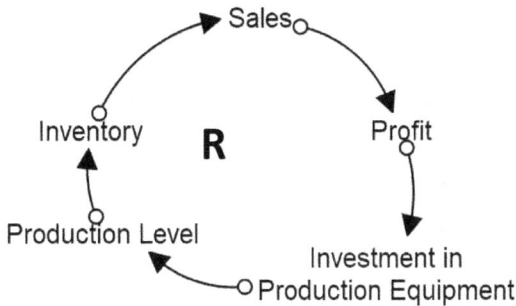

**Figure 21.** Sales Growth in a Capacity-Limited Production Environment.

*b) The Rules of Interaction*: The profit that is generated by sales is reinvested into additional production equipment, which increases the volume produced and thus the inventory of finished goods on hand. This allows an increase in sales. The assumption is that demand is so high that whatever is produced will be sold.

We stress that rules, policies, procedures, protocols, and laws, whether natural or man-made, and whether overtly articulated or unspoken, often explain the cause-and-effect relationships among system components; they therefore are very often components of systemic structure. In man-made systems, rules, policies, and procedures are often specified in company handbooks or similar policy manuals. But some rules remain unspoken, such as the common knowledge that you do not bother the Director on Monday mornings or that you do not arrive late to the V. P.'s meetings or that you do not take the last of the water in the cooler without replacing the jug. In natural systems, the natural "laws" exist whether they have been articulated by humans or not; they still govern cause-and-effect system component relationships.

## 4. Conclusions and Future Work

It seems that there is good agreement on the importance of "structure" in Systems Thinking, yet little agreement on the meaning of structure. Some experts argue that structure is the type or way that system components interact but do not specify what they mean by "type" or "way." Other researchers suggest that structure is the laws and rules governing system components interactions. Still others argue that structure is the physical components themselves. Most experts agree that structure is somehow involved with causality. We believe that in Systems Thinking, "structure" is the *cause-and-effect* manner in which system components interrelate to yield system behavior; and the rules, laws, protocols, procedures, policies, and incentives/rewards that govern those interactions. This definition has a sound theoretical basis and should prove useful to researchers, practitioners, and academics who are trying to both develop and understand systems. An empirical study validating this definition would be of great value to the Systems Thinking community.

## References

1. Monat, J. P., and Gannon, T.F., "What Is Systems Thinking? A Review of Selected Literature Plus Recommendations," *Am. J. of Systems Science,* **4**:2, 2015

2. Encyclopedia.com, Oxford University Press; https://www. encyclopedia.com/science-and-technology/computers-and-electrical-engineering/computers-and-computing/structure; accessed 14 December 2022

3. Webster's New Collegiate Dictionary, 1981, G & C Merriam Co., Springfield, MA, 1146

4. Cambridge English Dictionary, Cambridge University Press, 2022, https://dictionary.cambridge.org/us/dictionary/english/structure, accessed 14 December 2022

5. Dictionary.com, 2022, https://www.dictionary.com/browse/structure#:~:text=Definition%20of%20structure,building%2C%20bridge%2C%20or%20dam, accessed 14 December 2022

6. Collins Dictionary, CollinsDictionary.com, HarperCollins Publishers, Westerhill Road, Bishopbriggs, Glasgow, Scotland, G64 2QT, https://www.collinsdictionary.com/us/dictionary/english/structure; accessed 14 December 2022

7. Kim, Daniel H., *Introduction to Systems Thinking*, Pegasus Communications (now Leverage Networks Inc., www.leveragenetworks. com), 1999, ISBN 1-883823-34-X.

8. Senge, Peter, Kleiner, Art, Roberts, Charlotte, Ross, Richard, and Smith, Bryan, *The Fifth Discipline Fieldbook,*. Doubleday, New York, 1994.

9. Meadows, Donella H., *Thinking in Systems: A Primer*, Chelsea Green Publishing, White River Junction, VT, 2008, 89.

10. Stillwell, Doug, "Trust as a Systemic Structure in Our Organizations," The Systems Thinker, © 2018, Leverage Networks, Inc., https://thesystemsthinker.com/trust-as-a-systemic-structure-in-our-organizations/, accessed 25 October 2022

11. Spirkin, Alexander, "System and Structure," *Dialectical Materialism*, Chapter 2: The System of Categories in Philosophical Thought, © 1983 by Progress Publishers, Transcribed by Robert Cymbala, https://www.marxists.org/reference/archive/spirkin/works/dialectical-materialism/ch02-s07.html, accessed 15 October 2022

12. Karash, Richard, "How to See Structure," *The Systems Thinker*, 2018, (https://thesystemsthinker.com/how-to-see-structure/#:~:text=The%20essence%20of%20structure%20is,as%20structure%20is%20often%20hidden.)

13. Gharajedaghi, Jamshid, *Systems Thinking—Managing Chaos and Complexity: A Platform for Designing Business Architecture*, 2nd Ed., Elsevier, Burlington MA, 2006, 110.

14. Mcnaughton, Bruce, "System Structure (Pattern of Organization)," Enterprise as a System of Systems, Customer Driven Solutions

Limited©2022, https://eaasos.info/Content/EntSoS/Structure.htm, accessed November 28, 2022

15. Austin, Mark, "Modeling System Structure and System Behavior," ENES 489P Hands-On Systems Engineering Projects, Institute for Systems Research, University of Maryland, College Park, MD, https://user.eng.umd.edu/~austin/enes489p/lecture-slides/2012-MA-Behavior-and-Structure.pdf, accessed 15 December 2022

16. Barile, Sergio and Saviano, Marialuisa, "Foundations of Systems Thinking: The Structure-System Paradigm" (2011). in Various Authors, Contributions to Theoretical and Practical Advances in Management. A Viable Systems Approach (VSA). ASVSA, Associazione per la Ricerca sui Sistemi Vitali. International Printing, pp. 1-24, Available at SSRN: https://ssrn.com/abstract=2044579]

17. Anderson, Virginia, and Johnson, Laura, *Systems Thinking Basics From Concepts to Causal Loops*, Pegasus Communications, Inc., Cambridge, 1997.

18. Senge, Peter, *The Fifth Discipline*, Doubleday, New York, 1990; revised 2006, *52-53 and 286*.

19. Stroh, D. P., *Systems Thinking for Social Change*, Chelsea Green Publishing, White River Junction, VT, 2015, ISBN 978-1-60358-580-4

20. Cabrera, D. and Cabrera, L., *Systems Thinking Made Simple*, 2nd Ed., Plectica Publishing, Ithaca, NY, 2015, ISBN 978-1-948486-02-6

21. Cabrera, D.; Cabrera, L.; Cabrera, E.; "Relationships Organize Information in Mind and Nature: Empirical Findings of Action-Reaction Relationships (R) in Cognitive and Material Complexity," *Systems*, **10** (3), 2022, 71, https://doi.org/10.3390/systems1003007122. Lawson, Harold, *A Journey Through the Systems Landscape*, College Publications, London, 2010.

23. Maani, K. E., & Cavana, R. Y., *Systems Thinking, System Dynamic: Managing Change and Complexity* (2nd Ed.), Pearson: Prentice Hall, 2007.

24. Bellinger, Gene, "Systems Thinking – A Disciplined Approach," http://www.systems-thinking.org/stada/stada.htm.

25. Bellinger, Gene, "Translating Systems Thinking Diagrams to Stock & Flow Diagrams," 2004, http://www.systems-thinking.org/stsf/stsf.htm.

26. Calancie, L., Anderson, S., Branscomb, J., Apostolico, A. A., Lich, K. H. "Using Behavior Over Time Graphs to Spur Systems Thinking Among Public Health Practitioners," *Prev Chronic Dis* 2018; **15**:170254. DOI: http://dx.doi.org/10.5888/pcd15.170254external icon.

27. Hoehner, Christine M., Sabounchi, Nasim S., Brennan, Laura K., Hovmand, Peter, and Allison Kemner; "Behavior-Over-Time Graphs: Assessing Perceived Trends in Healthy Eating and Active Living Environments and Behaviors Across 49 Communities," *Journal of Public Health Management and Practice*, Vol. 21, Supplement 3: "Evaluation of the Healthy Kids, Healthy Communities National Program" (May/June 2015), pp. S45-S54, Lippincott Williams & Wilkins

28. Lannon, Colleen, "Structure-Behavior Pairs: A Starting Point for Problem Diagnosis," *The Systems Thinker*, 2018, https://thesystemsthinker.com/structure-behavior-pairs-a-starting-point-for-problem-diagnosis/ accessed 7 December 2022

29. LOC's Public Domain Archive, Cousin Cottage, Public domain, Library of Congress, http://www.loc.gov/rr/print/res/114_habs.html; https://loc.getarchive.net/media/cousin-cottage-1231-marais-street-faubourg-treme-new-orleans-orleans-parish-15)

30. 01 The Solar System PIA10231, mod02, Licensed Under Creative Commons Attribution 4.0 International CC by 4.0, Flickr. This image is modified from NASA's Photojournal Home Page graphic released in October 2007, catalog ID: PIA10231 (NASA/JPL), https://www.flickr.com/photos/11304375@N07/2818891443, accessed 17 October 2022.

# The Case for Systems Thinking in Undergraduate Engineering Education

Jamie Monat, Thomas Gannon, and Matthew Amissah
Systems Engineering Program, ECE Department
Worcester Polytechnic Institute
Worcester, MA 01609
USA

*Reprinted from International J. of Engineering Pedagogy, Vol. 12 No. 3, 31 May 2022, DOI:* https://doi.org/10.3991/ijep.v12i3.25035

*Abstract*— Systems Thinking is currently a popular topic among the general public as well as in colleges and universities. Systems Thinking helps solve complex problems that cannot be solved using conventional means, helps engineers design functional and reliable systems, and helps to understand why the world looks and behaves as it does. However, Systems Thinking is usually not integrated into engineering curricula; instead, it is either taught as a stand-alone, independent program or course, or it is not taught at all. We believe that this is sub-optimal. Just as mathematics, physics, and computer science are integrated into engineering courses as *fundamental tools,* so should Systems Thinking be. Here we provide several examples demonstrating the application and benefits of Systems Thinking to a variety of engineering disciplines with the objective of convincing university administrators and instructors to integrate it into every engineering discipline.

## 1. Introduction

The importance of understanding and dealing with engineering problems from a systems perspective is widely acknowledged. The Internet plus advances in computers and communication technologies have increased the interconnectivity among engineered products and systems, ancillary and support systems, infrastructure, society, and the environment. This has implications for engineering educational institutions' role in educating students on system (as well as component) interactions, feedback, and emergent properties. Many hiring managers now expect that freshly minted engineers will graduate with this awareness.

In industry, product manufacturers are interested in integrating information about component behavior with information from the product's stakeholder and operational domains. The expectation is that technological product engineers have a holistic appreciation of the product, informed by models from technical disciplines (such as mechanics, electronics, hydraulics, and thermodynamics) as well as by product behavior within an ecosystem of users and ancillary technologies. Today's engineers are expected to answer questions pertaining to product reliability, safety, economic viability, and sociological and environmental impacts.

As an example, the automobile industry is currently moving towards intelligent, autonomous, and sustainably powered vehicles. The current state of the art encompasses self-driving, electric powered vehicles capable of communicating with phones, manufacturers' servers, and other vehicles on the road [1]. Similarly, the current state of the art in Smart Homes entails networked appliances and utility services, with the capability for exchange of operational data and remote control by users and manufacturers [2].

Beyond the realm of industry, our critical infrastructure (*e.g.,* power systems, transportation systems, communications systems, and potable and wastewater systems) is undergoing a similar engineering paradigm shift. Increasingly the emphasis is on how to improve the performance of these systems by recognizing and exploiting the interconnections of the infrastructure with the various interfacing technological systems, society, and the environment. For example, with Smart Highways, cities will be able to better control traffic (especially in emergencies), as well as optimize road maintenance schedules. Smart grids will enable flexibility in harnessing energy by supporting input from various energy generation systems, as well as by optimizing power supply and distribution based on changing demand.

Humanity may also be viewed as a system [3]. In the past, whereas psychology, sociology, culture, politics, and economics were viewed as domains separate from and independent of engineering, it is now recognized that these societal elements are intimately tied to the technologies that serve them. Considering technology in isolation from them is foolish. Technologies fail where cultures do not embrace them; cultures fail where they do not adopt technologies that their competitors do. Today's engineers must understand the psychological, cultural, economic, political, and environmental implications of their engineering decisions.

Traditional engineering education does not provide the broad systemic perspective outlined above, nor does it facilitate holistic thinking. Conventional analytic tools such as differential equations are useful but often do not yield insights into complex problems, and they are sometimes unwieldy or unsolvable in closed form. Basic equations, design paradigms, standard procedures, and rules of thumb have served us well in the past but fall short as the world becomes more interconnected and as solutions become less black-and-white.

Therefore, the current state and trend of technology highlight the need for a systems perspective in engineering. Engineers of today must be trained to think and problem-solve as Systems Thinkers. They must be able to appreciate all the factors (not just traditional engineering factors) that come to bear in real-world problems. They must also recognize that there is usually more than one solution to any complex problem and that their engineering judgment must be tempered with an appreciation for sociological, cultural, ethical, and psychological factors. Addressing socio-technical challenges such as energy, food security, health, poverty, sustainability, and clean water requires a global perspective: perceptions of engineering solutions acceptable to various cultures can be very different, and globally aware engineers need to understand and anticipate these perception differences to make an even greater impact on pressing global challenges.

ABET (the organization in charge of accrediting science, computing, technology, and engineering college and university programs) lists 11 outcomes [4] that all engineering baccalaureate graduates should possess. Of these we identify the following as essential Systems Thinking skills:

- "an ability to identify, formulate, and solve complex engineering problems by applying principles of engineering, science, and mathematics" (3.a)

- "an ability to apply engineering design to produce solutions that meet specified needs with consideration of public health, safety, and welfare, as well as global, cultural, social, environmental, and economic factors" (3.b) [4]

This raises the question of how to effectively integrate Systems Thinking into engineering education. In this paper we argue that Systems Thinking entails concepts, methods, and tools that should be taught in traditional engineering courses to engender a more holistic perspective in students.

There are several reasons why Systems Thinking is not currently fully integrated into engineering education:

- Its methodology and tools are not well-understood by instructors.

- Its benefits are not understood.

- It is perceived to be a substitute or replacement for (as opposed to an adjunct to) conventional, tried-and-true engineering approaches.

- Engineering curricula are full and cannot accommodate additional courses.

We propose an approach for exploring Systems Thinking concepts and tools based on the major themes of 1. Conceptual modeling 2. Dynamic modeling and 3. Holistic thinking tools. We offer several examples illustrating implementations of these ideas in the context of typical engineering problems.

Subsequent sections of the paper are organized as follows: In Section 2 we provide a literature review; in Section 3 we offer an overview of Systems Thinking highlighting important concepts and principles applicable to engineering. In Section 4 we discuss the application of systems thinking tools to typical engineering problems, demonstrating how these may be integrated into existing curricula to support a broader analysis framework. Section 5 presents the results of a preliminary field study. In Section 6 we offer a summary of our proposed solutions for integrating Systems Thinking. Section 7 concludes with a summary of expected contributions and impact of Systems Thinking in engineering education.

## 2. Literature Review

Several research publications [5-7] have called for the incorporation of Systems Thinking into engineering education. Although there isn't a clear literature consensus on what Systems Thinking specifically entails, the expected learning outcomes are consistent: to equip students with competencies required to work collaboratively with others, to draw on multidisciplinary expertise to properly frame problems, and to devise technical solutions that adequately factor in the complexities of the operating environment and human interaction. These competencies are typically associated with the disciplines of Systems Thinking, Systems Engineering, and Systems Design. Contributions in the literature include an identification of required Knowledge, Skills, and Abilities (KSA's); the description of essential topics and courses that have been implemented; and the results of surveys used to assess the effectiveness of such courses.

Hadgraft et al. [5] note that in Australia teaching Systems Thinking is already mandated by Australian accreditation guidelines and that it must include the contextual framework of social, cultural, ethical, legal, political, economic, environmental, sustainable, and safety considerations, which are very similar to ABET accreditation guidelines in the U. S. Hadgraft also notes that industry ranks Systems Thinking as important for engineering design and management work and touts it as a *core engineering competence*. They surveyed 307 chemical engineering and civil engineering undergraduates for their opinions and found that Systems Thinking skills are valued by the majority (77%) of students, but that only ~33% of the students felt that Systems Thinking skills were taught or assessed well.

Davidz et al. [6] conducted a field study to determine the enablers, barriers, and precursors to Systems Thinking development in engineers. Their study focused on 205 aerospace industry employees with various levels of experience and tenure. They concluded that the primary enablers of Systems Thinking were experiential learning, individual characteristics, and a supporting environment. However, their analysis focused on practicing Systems Engineers as opposed to academic curricula or content. It is interesting that their survey responses indicated that in 2007, education was *not* considered a principal enabler of Systems Thinking in engineers. They offer the following suggestions for academia: 1) Offer Systems programs 2) Use feedback mechanisms to continually improve Systems programs and courses 3) Structure courses to emphasize experiential learning 4) Structure courses to emphasize context and knowledge integration and 5) Continue research on the mechanisms for effective Systems Thinking development.

Muci-Kuchler *et-al.* [7] note that student teams in undergraduate Mechanical Engineering courses often struggle when they have to design products with several components requiring multiple areas of technical expertise, due to the watering down of typical design problems to remove the complexity that is typically present in practical systems. Valenti [8] similarly underscores a need for reforms in mechanical engineering curricula to better meet the demands of industry. His proposed Systems Thinking topics to address include a Systems Perspective, Teamwork, Communication, Creative Thinking, and Professional Ethics.

Based on the U.S. Department of Labor's Engineering Competency model [9], the Systems Engineering Career Competency Model (SECCM), [10] and the CDIO syllabus [11], Muci-Kuchler *et al.* offer a more extensive list of topics aimed at basic Systems Thinking and Systems Engineering KSA's. The primary topics identified include:

- System, Element and System boundaries
- System context and System of Systems
- System function, behavior and emergent properties

- System structure and decomposition

- System life cycle

- Basic types of system architecture

- Identifying Stakeholders

- Interfaces, interactions and dependencies

A typical undergraduate product design course was updated based on the aforementioned topics to include lessons in Systems Thinking and Systems Architecture. The emphasis here however was on Systems Engineering, and the subject of dynamic modeling (which is central to Systems Thinking) was not explored in detail.

A similar effort is offered in [12]. Here the required content was covered in two courses: an undergraduate and a graduate level course on *Systems Engineering and Design,* targeted for Industrial Engineering students. Students were first taught 'Classical Systems Engineering' topics, then later exposed to Sociotechnical System Theory (SST), Cognitive Systems Engineering (CSE) and Soft Systems Methodology (SSM). These methodologies were treated as complementary and useful in resolving a range of problems from "hard" to "soft" where all problem situations are presented on a continuum from well-structured to unstructured. Topics covered in the courses include: Design Options, Engineering Systems Modelling, Analysis of System Reliability, System Dynamics and State Transition Matrix Models, System Lifecycle, Optimization, Management of Engineering Systems Design and Operation, Introduction to Sociotechnical Systems Theory (SST), Introduction to Cognitive Systems Engineering (CSE), and Soft Systems Methodology (SSM). This work offers an extensive coverage of the main ideas in the field of Systems Thinking, and it may serve as a useful template for a Minor in the field. However, the subject of Systems Thinking is treated separately, rather than being integrated into existing engineering courses.

Rehman [13] describes the objectives of the *E2020 Scholars Program* at Iowa State University, which sought to have students become proficient in four pillar areas: leadership, innovation, global awareness, and systems thinking. Each pillar was introduced during three weeks in a freshman-level seminar followed by half of a semester in a year-long sophomore-level seminar. Students applied systems thinking to grand challenge problems by considering factors inside and outside of engineering and using three graphical tools. They identified connections between elements with rich pictures, explained relationships with causal loop diagrams, and sketched the behavior over time of key variables in the system. Qualitative observations and quantitative assessments suggested that the initial offerings were mostly successful. Most students stated that the activities helped them to appreciate the range of issues affecting an engineering problem. Students struggled most with identifying key variables and deriving the behavior over time from causal loop diagrams. This work offers some useful resources for implementing courses that offer an introduction and overview to Systems Thinking.

Degen *et al.* [14] provide an overview of a new course that was developed to help sophomore students in mechanical engineering develop skills in systems thinking. They provide details about an Engineering Systems Thinking Survey (ESTS) that was developed to assess systems thinking skills in specific areas and provide the results of the ESTS from implementation of the course during two separate semesters. The specific areas targeted by that survey were the identification of customer needs, establishing target product specifications, concept generation, and systems architecture. The survey results showed that the course was successful in improving students' self-efficacy on each of the four topics, particularly in setting target specifications and systems architecture. Comparisons of pre- and post-survey results showed improvements in student answers on the technical questions related to identification of customer needs, establishing target product specifications, and concept generation, with a slight decrease in the area of systems architecture. The survey results also indicated the need to strengthen students' awareness of concept implementation. Similar to the work by Muci-Kuchler *et al.*, this work is more pertinent to Systems Engineering, and the subject of dynamic modeling is not explored in detail.

Robinson-Bryant [15] describes a Systems Thinking skills intervention developed for an online Project Management course for 3rd and 4th year engineering students. It describes the application of a vertical course thread approach, called a Conceptual Systems Thinking Integration approach, which outlines instructional events, learning events, knowledge features, and assessment events that can be applied to facilitate "robust" learning of Systems Thinking skills. It also provides a literature-based discussion of the growing importance of developing an orientation towards Systems Thinking skills for all engineers. Similarly, in [16] Cattano *et al.* describe their approach in teaching Systems Thinking to Civil Engineering students at Clemson University in 2009. They elaborate on teaching methodologies: which specific exercises, assessments, and lecture topics worked best. Thus, these papers are a good initial reference for *how to teach* Systems Thinking to engineers.

Hung *et al.* [17] describe a case study in which they assessed the ability of eight students to apply Systems Thinking concepts both before and after the students had taken a one semester modelling course. They argue that systems modelling is a *cognitive tool* that helps students understand Systems Thinking concepts. They found a statistically significant increase in students' utilization of systems thinking through interrelationships, causal relationships, and feedback processes, and concluded that teaching System Dynamic models is effective in increasing students' knowledge of and ability to apply Systems Thinking concepts. They mention several models that were used (population, thermostat regulating room temperature, growth of a city on limited land, prey and predator) but they do not discuss the specific details of the models.

Jain *et al.* [18] note the need to include societal, business, and environmental considerations in engineering designs, and they argue that Systems Thinking should be part of the core undergraduate engineering curriculum, but their paper is more about Systems Engineering than Systems Thinking. They also argue for starting Systems Thinking training in grades K-12.

Wasson [19] argues that there is an educational void in Systems Engineering instruction for engineers and bemoans the facts that Systems Engineering is taught primarily at the graduate (not undergraduate) level, that its acceptance is limited by curriculum requirements, and that it is not introduced at the K-12 level. He suggests that a Systems Engineering undergraduate course be required for all engineering curricula and that we execute a paradigm shift to a Systems Engineering methodology-based education. Like Jain, Wasson refers primarily to Systems Engineering, not Systems Thinking.

The growing need for inculcating Systems Thinking into undergraduate and graduate engineering curricula is also being highlighted by industry. For example, Summerton et al. [20] note that Systems Thinking facilitates a more integrated understanding of related subject matter, as opposed to teaching disparate concepts. They claim that Systems Thinking improves student learning by promoting the consideration of a wide range of both positive and negative impacts within the context of multiple interacting systems, and Systems Thinking allows students to make predictions based on their understanding of how system outputs may change given changes in an input or parameter. They also observe that Systems Thinking prepares students more thoroughly for their future career paths.

Voorhees and Hutchison [21] state that in the Green Chemistry industry *"industrial employers are looking to hire students with expertise in sustainable practices and an understanding of systems thinking and life cycle."* In the lean manufacturing industry, Ballé [22] observed from his personal experience that his work with Systems Thinking concepts and System Dynamics simulations prior to studying lean practices facilitated his immediate application of lean practices, which was not the case for his colleagues that did not have a background in Systems Thinking.

The literature clearly underscores the need for and benefits of Systems Thinking in engineering education. However, there is significant diversity in the topics taught, ranging from Systems Engineering and

Systems Architecture to Soft Systems Methodology and Critical Systems Thinking. Outside of proposing new courses that address these topics independently, we hope to contribute to this discussion by demonstrating that Systems Thinking can be integrated into most existing engineering courses. We propose that incorporating systems modeling approaches into existing courses exposes students to practical analysis tools that are flexible enough to support problems that are ill-structured, incomplete, and ambiguous. Additionally, we propose that existing project-based course work can be extended to support interdisciplinary collaboration and to address environmental and socio-cultural issues.

## 3. Systems Thinking Overview and Tools

Systems Thinking has been characterized as a perspective, a language, and a set of tools [23]. It is a holistic perspective that recognizes that the relationships among system components and between the components and the environment are as important as the components themselves. It is a language of feedback loops, emergence, complexity, hierarchies, self-organization, dynamics, and unintended consequences. Systems Thinking tools include the Iceberg model, causal loop diagrams, behavior-over-time plots, stock-and-flow diagrams, systemic root cause analysis, dynamic modeling tools, and archetypes. See [23] for a more comprehensive explanation of Systems Thinking.

We distinguish between Systems Thinking and Systems Engineering, which is an emerging discipline and profession that focuses on the successful engineering of complex man-made systems [24]. Systems Engineering entails a systematic process of translating stakeholder needs into increasingly detailed design specifications, ultimately leading to the realization of physical/cyber-physical systems that are deployed, operated, and retired in time, in accordance with stakeholder expectations.

In our teaching, we have found it convenient to split the Systems Thinking tools into 3 categories:

1. Conceptual Modeling Tools to articulate and frame issues, elicit knowledge and beliefs, and meaningfully organize information to appreciate underlying causal structures.

2. Dynamic Modeling Tools to assess the dynamics of those causal structures, and to evaluate potential interventions.

3. Holistic Thinking Tools to ensure that complex problems are not addressed using unidimensional solutions.

### 3.1. Conceptual Modeling Tools

Robinson [25] defines Conceptual Modeling as a non-software specific description of a simulation model that is to be developed, describing the objectives, inputs, outputs, content, assumptions and simplifications of the model. This definition is mostly suited to the context of simulation development. In this paper we apply the term more loosely, from the perspective that Conceptual Modeling may not necessarily lead to a simulation model. Conceptual Modeling entails framing a problem precisely enough to allow for more rigorous analysis, but also transparently enough to engage a broad base of stakeholders and capture their perspectives on the problem. This should ideally inform a common baseline of what the problem entails, deepen stakeholders' understanding of the problem and its context, and support further analysis to inform how the problem should be addressed.

Conceptual Modeling tools offer a language to describe issues in general concepts that are accessible to a broad range of stakeholders, while also abstracting the details of constructs required for dynamic analysis. For example, constructs such as Stocks, Flows, and Causal Loops offer a simple way to visualize the structure and dynamics of systemic problems ranging from machine control to the spread of disease.

There are several Conceptual Modeling approaches incorporating tools such as general graphical modeling languages, simulation framework-specific languages, Behavior-Over-Time plots, Stock and Flow diagrams, Causal Loop Diagrams, the Agents Modeling Language (AML), and the Iceberg Model.

### 3.2. Dynamic Modeling Tools

Dynamic models may be mathematical or simulation models, used to assess how a system changes over time. These allow us to exercise and test our assumptions, hypotheses, and knowledge of a system especially in situations where experimentation on the actual system is infeasible and/or prohibitively expensive. Systems Thinking heavily emphasizes understanding the dynamics of systems before any interventions are made, as it is often true that the problems of today are the result of the fixes of yesterday.

Differential equations are currently the dominant dynamic modeling approach in engineering. The power and limits of calculus, however, must be weighed carefully. While calculus is rigorous and effective for making generalizations about system behavior, it is an arcane language, unintuitive to most, and often too restrictive in the nature of problems that can be modeled. Furthermore, some differential equations are just unsolvable in closed form. Given the emphasis on dynamics in Systems Thinking we argue for the adoption of simulation/computational models as adjuncts to (not replacements for) differential equations. These are more expressive, flexible, and applicable to a broader range of problem contexts, albeit with some limitations in the generalization power.

There are several simulation methodologies available. System Dynamics [26] is the predominant simulation methodology in Systems Thinking. According to Forrester [27] it is a necessary foundation underlying effective thinking about systems. Recently the field has embraced a broader perspective on dynamic modeling tools including other frameworks such as Cellular Automata, Discrete Event Simulation, and Agent Based Modeling.

### 3.3. Holistic Thinking Tools

Politicians, businesspeople, and bureaucrats often attempt to solve complex problems unidimensionally: that is, by throwing money at the problem. An example is the 2008 U. S. bank bailout after the financial/housing crisis. These unidimensional "solutions" sometimes address short-term symptoms, but often do not solve the underlying causes and may lead to other problems. The 2008 bailout, for example, led to the enrichment of bank executives who had made poor decisions and created incentives for managers to take unreasonable risks with the knowledge that they will be bailed out by taxpayer dollars; it also retarded economic development by rewarding failure.

Technocrats and engineers make a similar mistake by assuming that complex problems such as poverty, hunger, terrorism, climate change, pollution, and the lack of potable water can be solved by technology alone. But technology alone does not address the political, economic, environmental, ethical, psychological, and cultural aspects of the situation. In many cases, these non-technical aspects dominate system performance. For example, in the 2014 Ebola outbreak in Guinea, Liberia, and Sierra Leone, the control strategy involved isolating and avoiding contact with sick people. However, the West African culture involved strong family values, which mandated that the ill be cared for by family members who also wash the bodies of their dead [28]. These traditions proved to be major obstacles to the strictly technical solution.

Another example involves the Amish, whose traditions eschew motorized vehicles, the use of electricity, and the taking of photographs. Electric vehicles would not be successful in that culture. Yet another example involves vaccination. Many vaccines are stabilized with pork-derived gelatin. Orthodox Jews and conservative Muslims cannot use pork products, so conventional vaccines are not suitable for them.

Engineers must therefore be trained to think holistically. Specific Systems Thinking tools that facilitate holistic thinking include System Breakdown Structures, System Interrelationship Matrices, and Causal Loop Diagrams. But a well-written paragraph may be all that is needed. Also, the ability to identify and understand worldviews and mental models is a skill critical to holistic thinking. This capacity is facilitated by tools (such as the Iceberg Model) that help discover, expose, assess, and revise our ubiquitous (and often incorrect) Mental Models.

# 4. Descriptions and Examples of Systems Thinking Tools for Engineering

In this section we elaborate on specific Systems Thinking tools discussed in the previous section and provide examples of their application. The aim is to show how these tools offer a complementary approach for framing and understanding typical engineering problems and how to apply them to address complex societal problems.

## 4.1. Conceptual Modeling Overview and Application

*4.1.1. Causal Loop Diagrams.* According to Monat and Gannon [29] one of the first steps in attempting to understand system behavior is the construction of a causal loop diagram, which graphically portrays systemic causes and effects and helps to identify reinforcing and balancing feedback loops. A simple temperature control causal loop diagram for heating is shown in Figure 1:

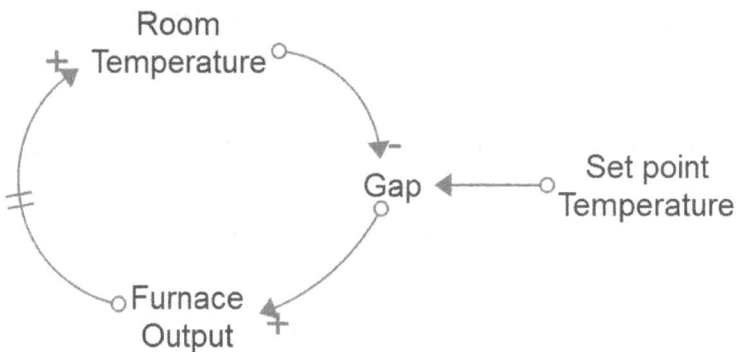

Figure 1. Temperature Control Causal Loop Diagram

*4.1.2. Stock & Flow Diagrams.* Systems usually require the storage or accumulation of things, which may include physical quantities such as the volume of liquid, quantity of electric charge, number of deer in a field, number of clients of a company, or amount of money in a bank account. Those things can also be non-physical things, which could include emotions such as love, greed, anger, or lust. These quantities of things in systems are called stocks, which can increase or decrease due to flows into or out of them. Stock and flow diagrams illustrate the stocks, inflows, and outflows of things in a system. These diagrams are usually developed in concert with causal loop diagrams and are important first steps in system dynamics modeling. Stock and flow diagrams, along with causal loop diagrams, provide valuable insights in understanding system behavior. A simple stock-and-flow diagram showing logging impact on a forest [30] is shown below in Figure 2.

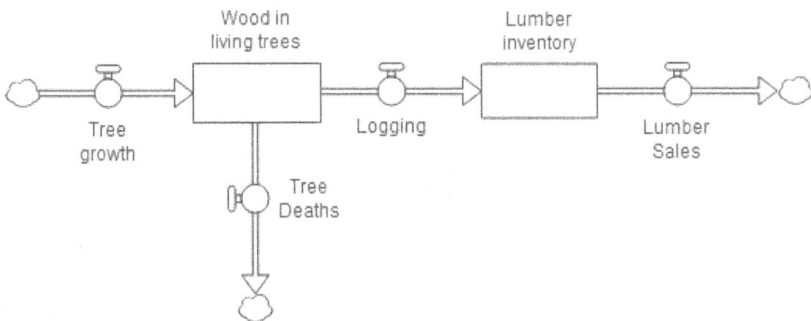

Figure 2. A Basic Stock-and-Flow Diagram [30]

4.1.3. *Behavior-Over-Time Plots.* When one first comes upon a system, it can be very difficult to understand the way the system works or the systemic structure. Behavior-Over-Time (BOT) plots are therefore useful. BOT plots simply show the value of one or more system parameters over time. Several examples are shown in Figure 3.

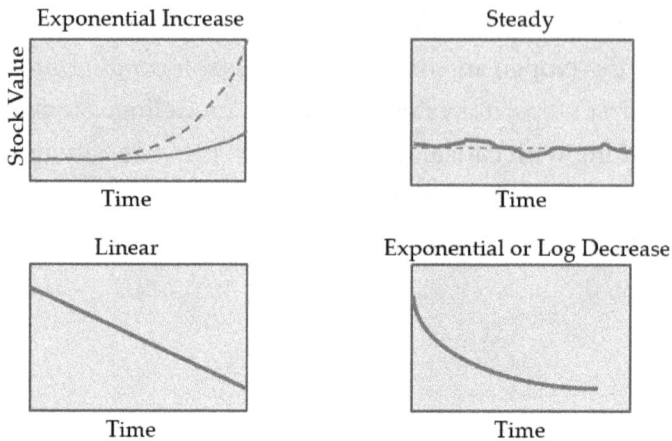

Figure 3. Behavior-Over-Time Plots

A parameter that increases or decreases exponentially indicates the presence of a reinforcing feedback loop. A parameter that oscillates indicates the presence of feedback loop with delays. A linear parameter indicates either the absence of feedback loops or broken feedback loops. A parameter that remains constant over time indicates the presence of a stabilizing feedback loop. Thus, simple observation of BOT plots provides insight into the systemic structure.

*4.1.4. The Iceberg Model.* The Iceberg Model is a convenient uber-tool for understanding the systemic big picture. It posits that repeated observable events are patterns, that patterns are caused by systemic structure (hierarchies and feedback loops), and that structure is caused by underlying forces (mental models in human-designed systems; natural forces such as gravity and electromagnetism in natural systems.) The Iceberg Model is depicted in Figure 4.

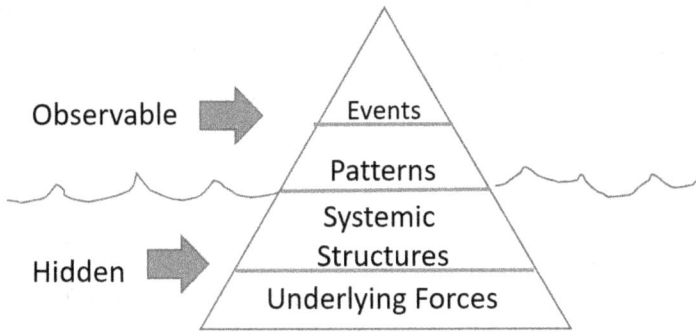

Figure 4. The Iceberg Model

Events and patterns are typically observable while the underlying systemic structures and mental models are not and must be uncovered.

*4.2. Systems Dynamics Overview and Application*

System Dynamics was developed in the 1950's and 1960's, mostly through the work of M.I.T.'s Jay Forrester who adapted the mathematics of control theory to the dynamic modeling of business decisions and subsequently to urban and global policy analysis [31-33]. According to Forrester, System Dynamics involves interpreting real life systems as computer simulation models that allow one to see how the structure and decision-making policies in a system create its behavior.

Monat and Gannon [29] state, "In their most basic form, System Dynamic models are typically control volume analyses: an initial quantity or stock increases over time due to an inflow and decreases due to an outflow" as shown in Figure 5:

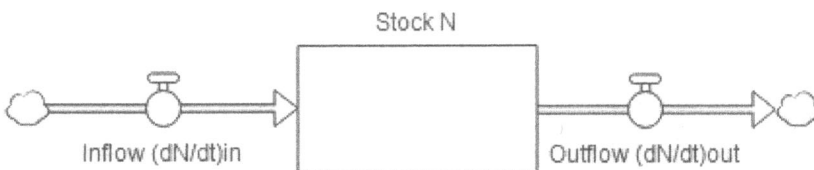

Figure 5. Control Volume Analysis

In the above example, $t$ is time and $dN/dt$ represents the instantaneous change in the quantity of stock $N$ with respect to time. Given that $N_0$ represents the initial value of the stock, this model implements equation 1, which is used to calculate the population N as the simulation advances in time increments $\Delta t$.

$$N(t) = N_0 + [(dN/dt)_{in} - (dN/dt)_{out}]\Delta t \qquad (1)$$

Modeling the dynamics of a system usually starts with developing a causal loop diagram and then translating into a stock and flow diagram. Next, links between stocks and flows are added along with initial values for each stock. Algebraic equations are then developed to quantify the inflows into and outflows from the stocks, and a simulation is run and debugged. Behavior-Over-Time plots are usually used to display the results, which must then be compared with reality to validate the model. After the results are validated, control points may be identified, and experiments conducted to see how to best influence the system. Detailed instructions on how to model the dynamics of a system are provided by Barry Richmond [34].

In the following subsections we apply Systems Dynamics to several engineering problems that are traditionally solved using differential equations. We argue that Systems Dynamics is especially well-suited to such problems due to its flexibility in adjusting model parameters and intuitive graphical language supporting conceptual modeling. This approach scales better for real world dynamic analysis and should be considered as a valuable addition to traditional differential equations for teaching dynamic engineering analysis.

*4.2.1. Radioactive decay.* A typical decay or depletion process can be described by the simple Causal Loop Diagram (CLD) shown in Figure 6.

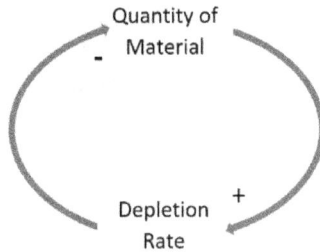

Figure 6. Causal Loop Diagram for Depletion

The CLD indicates that a greater quantity of material increases the depletion rate but that a higher depletion rate reduces the quantity of material. Figure 7 shows the corresponding stock-and-flow diagram indicating that the rate of uranium decay is proportional to the quantity of uranium remaining. Once this schematic is entered into the Stella Architect software, one need only enter an initial concentration for the starting stock of uranium and that the decay rate is equal to a constant time the concentration of uranium per the equation $dN/dt = -kN$ (where $N$ is the concentration of a radioactive species such as $U$ 238, $t$ is time, and $k$ is the radioactive decay constant); the model then yields the familiar exponential decay curve shown in Figure 8.

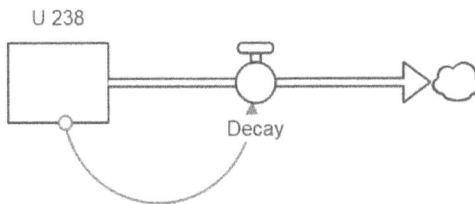

Figure 7. Stock-and-Flow for U-238 spontaneous decay

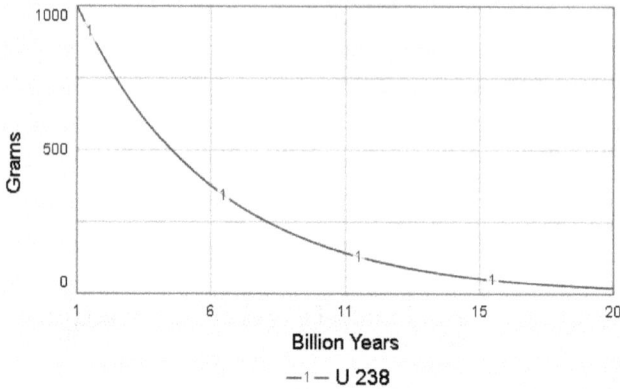

Figure 8. Radioactive Decay

A conventional engineering solution to this problem would require the integration of the equation $dN/dt = -kN$ which is far less descriptive and intuitive than the System Dynamic perspective.

*4.2.2. Fluid Mechanics: Tank Drainage.* A simple tank drainage model exemplifies the weakness in conventional engineering education and the benefits of Systems Thinking. Figure 9 shows a cylindrical 5000-gallon tank filled with water. At time zero, a valve near the bottom of the tank is opened, allowing the water to gush out. What is the volume of water in the tank as a function of time?

V gallons of water in the tank

Drainage Rate $dV/dt$

Figure 9. Tank Draining

A conventional solution to this problem would argue that the drainage rate, $dV/dt$ equals some constant $k$ times the instantaneous volume of water in the tank $V$, i.e. $\frac{dV}{dt} = -kV$. Solving this by rearranging terms and integrating yields the exponential relationship $V = V_0 e^{-kt}$. Where $Vo$ is the initial volume of water in the tank. A plot of the water volume vs time is shown in Figure 10.

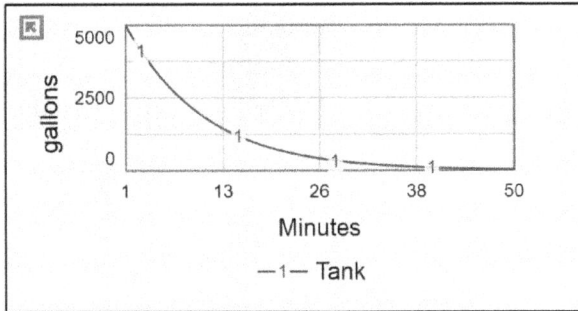

Figure 10. Tank Volume vs Time

Yet in our introductory course on Systems Thinking, many engineering students get this plot wrong, drawing a linear decay of volume over time. When this occurs, we ask the students to consider a smaller cylindrical tank on a tabletop, as shown in Figure 11:

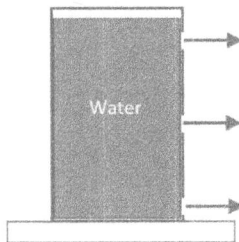

Figure 11. Tank on Table

We ask the students to imagine 3 small holes drilled into the side of the tank at various elevations and ask the students to sketch the trajectories of the water exiting from the 3 holes. They eventually figure out that the water spurts vigorously out of the bottom hole, less so out of the middle hole, and barely trickles out of the top hole. When we ask why this is so, they reply that the pressure increases toward the bottom of the tank because of the greater volume of water pressing down due to gravity. We then take them back to the draining 5,000-gallon tank and ask them to re-plot the tank volume vs time, and this time they get it right. Systems Thinking provides an interactive conceptualization that allows a student to visualize the dynamics of a situation as opposed to a differential equation which often leaves students stuck with the mathematical mechanics of the problem. A stock-and-flow diagram for this situation is shown in Figure 12 and the corresponding Causal Loop Diagram is provided above in Figure 6. The System Dynamic model output is shown above in Figure 10. The stock-and-flow diagram indicates that the rate of water draining is proportional to the volume of water remaining in the tank.

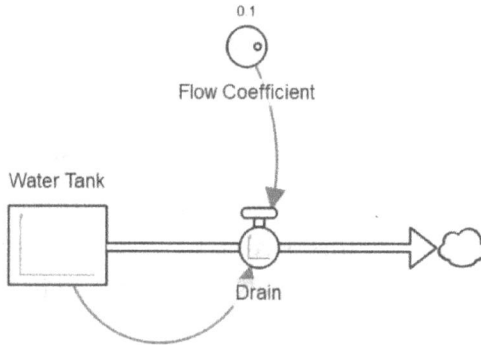

Figure 12. Drainage Stock-and-Flow Diagram

Here again the System Dynamics approach provides insights not afforded by the conventional differential equation solution.

*4.2.3. Heat Transfer: Ice Freezing on a Pond.* In winter, how fast does ice build up on the surface of a pond, and when will it be thick enough to walk on safely? Consider the schematic shown in Figure 13 in which a newly forming volume of ice of area A and thickness dx is forming just under the surface of existing ice on a pond.

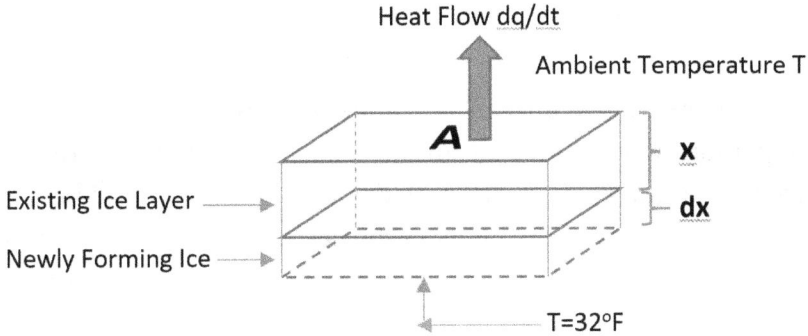

Figure 13. Schematic of Ice Freezing at the Surface of a Pond

Define:

$\rho$ = density of ice = 0.036 lbm/cubic inch

$\Delta H_f$ = Water heat of fusion = 143 BTU/lbm

$k$ = thermal conductivity of ice = 2.3 BTU/(day-inch-degree F)

$x$ = thickness of ice in inches

$A$ = area in square inches

$t$ = time in days

$T$ = temperature in degrees F

$dq/dt$ = heat flow in BTUs/day

The rate of heat flow through the ice, $dq/dt$ is provided by the standard conductive heat flow equation:

$$dq/dt = kA(32 - T)/x$$

And $dq$, the amount of heat transferred in time dt is then

$$dq = kA(32 - T)dt/x \qquad (2)$$

The amount of heat that must be extracted to freeze a volume of water with dimensions $A \cdot dx$ is given by

$A \cdot dx \cdot \rho \cdot \Delta H_f$. Equating this to $dq$ in equation 2 yields

$$\rho \, \Delta H_f \, A \, dx = kA(32 - T)dt/x$$

Which may be rearranged and simplified to yield

$$dx/dt = k(32 - T)/(x \, \rho \, \Delta H_f) \qquad (3)$$

which is the rate of ice growth over time. A conventional solution to this equation would involve multiplying both sides by dt and integrating to obtain a square root dependence of ice thickness on time. But System Dynamics obviates the need for integration. Instead, one may draw a very simple stock-and-flow diagram for this situation as shown in Figure 14.

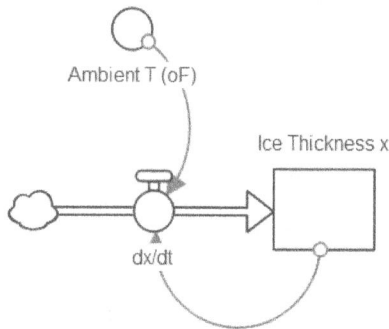

Figure 14. Stock and Flow Diagram for Ice Freezing on Pond Surface

Then all one must do is enter $k(32 - T)/(x \, \rho \, \Delta H_f)$ for $dx/dt$ in the Stella Architect model and run the model, plotting x, the ice thickness over time. The result is the graph shown in Figure 15.

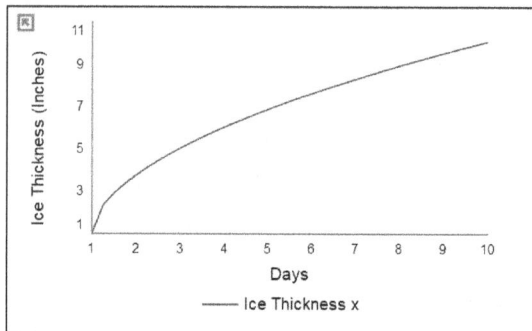

Figure 15. Ice Thickness vs Time

### 4.2.4. Mechanical Engineering: Simple Harmonic Motion. A simple harmonic oscillator subject to both gravitational and spring forces is depicted in Figure 16.

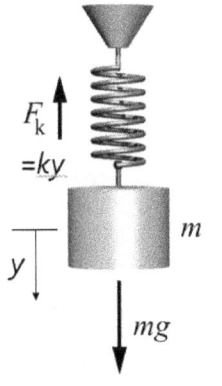

Figure 16. Spring-Mass System

Traditional methods of solving for the mass's position as a function of time involve guessing a solution of the form: $y = Ae^{i\omega t}$. Where $y =$ the mass's vertical position, $i = \sqrt{-1}$, $k =$ the spring constant, $t =$ time, $m =$ the mass of the oscillator, and $\omega = (k/m)^{.5}$. The expression is then differentiated twice and substituted into the expression $F = m\ddot{y} = mg - ky$, the imaginary term is disregarded, and one obtains the solution $y = A\cos[(k/m)^{.5} t]$. This method of solution is not intuitive to most students, and it is hard to understand why a solution must be guessed, as well as what the guess should be. In addition, since the solution is not analytic, might there not be other solutions that also work?

System Dynamics provides an alternative (and perhaps more intuitive) methodology. First, a Causal Loop Diagram (Figure 17) is drawn:

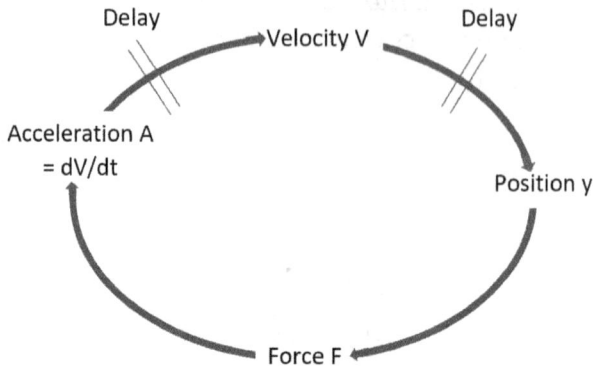

Figure 17. Spring-Mass Causal Loop Diagram

The mass's vertical position y determines the magnitude of the force that acts upon it while simultaneously the force on the mass impacts the mass's acceleration which impacts the mass's velocity which impacts the mass's vertical position y. However, while the position instantaneously changes the force, the force does not instantaneously affect the mass's position; there is a delay as the force produces an acceleration that changes the mass's velocity and, over time, its position. It is this delay that causes the oscillation. This is a rule of thumb for Systems Thinking: feedback loops with delays typically cause oscillation. Therefore, if one observes an oscillation in nature, it behooves one to figure out the systemic relationships and identify the causative feedback loops. A stock-and-flow diagram (Figure 18) may then be drawn for this situation:

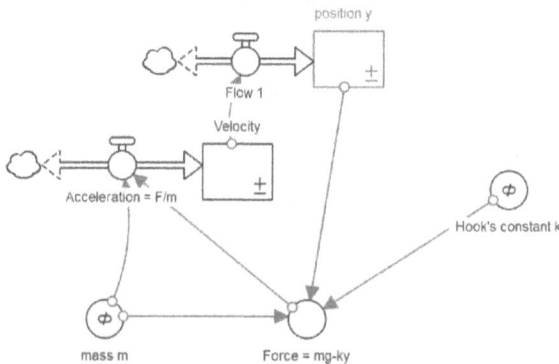

Figure 18. Spring-Mass Stock-and-Flow Diagram

The right side of this model shows the Hooke's Law force (the spring force) resulting from spring compression and gravitational force: $F = mg - ky$. But also, $F = ma$, so the spring and gravitational forces cause an acceleration of the mass m, which is shown on the left side. The force responds instantaneously to the position and so does the acceleration, since $F = ma$. But the acceleration causes a velocity change over time (velocity does not respond instantaneously to acceleration), and the velocity causes a position change over time (position does not respond instantaneously to velocity.)

A sample set of parameters for this model could be:

Initial Position y = 50 mm

Hooke's Constant k = 0.5 N/mm

Force = (-1)(Hooke's Constant k)(Position y) + (9.8)(Mass m) (N)

Mass m = 1 kg

Acceleration = (-k)(position y)/(mass m)        $(m/s^2)$

Initial Velocity = 0 m/s

Flow 1 = Velocity m/s

This model was entered into isee Systems' Stella Architect system dynamics modeling software, and yielded the following plot (Figure 19) of the mass's position and velocity vs time:

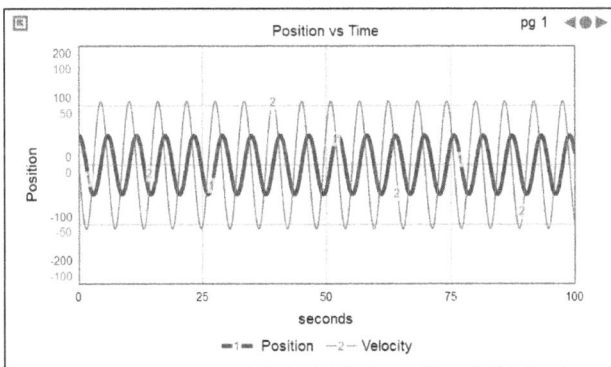

Figure 19. Stella Model Output

89

This is an intuitive, simple method of solving for the position of the mass over time involving only fundamental concepts of force, mass, acceleration, and velocity.

4.2.5. *Electrical Engineering: Oscillating Circuits.* LC circuits, like the one depicted in Figure 20, are common in radio, TV, tuners, oscillators, signal generators, mobile phones, power systems, and electronic filters. In the figure, V is voltage, L is inductance, and C is capacitance.

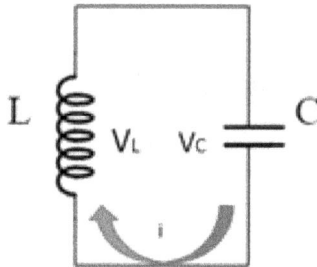

Figure 20. Simple LC Circuit

Traditionally, the current flow over time in these circuits is solved using differential equations analogous to those used for the simple harmonic mechanical oscillator described above. Defining V= voltage, q = charge, i = current, and t = time, we have:

$$V_C = -q/C \qquad (4)$$
$$V_L = L\,di/dt \qquad (5)$$
$$di/dt = V/L \qquad (6)$$
$$i = (di/dt)dt \qquad (7)$$
$$q = idt \qquad (8)$$

The conventional solution for the current flow in this circuit is quite involved, involving a second-order differential equation for which we guess a solution of the form $i(t) = Ke^{st}$ where $i(t)$ is the current, $K$ and $s$ are constants, and $t$ = time. This is then differentiated twice and substituted into the Kirchoff's law equation $V_l + V_c = 0$, leading to a solution of the form: $i(t) = K_1 e^{j(1/LC)^{0.5}t} + K_2 e^{-j(1/LC)^{0.5}t}$ where $j = \sqrt{-1}$. Euler's formula is then used to replace the exponential terms with trigonometric terms, the imaginary terms are disregarded, and one obtains the solution: $i(t) = (C/L)^{0.5}V_o sin[(1/LC)^{0.5}t]$ where $V_o$ is the voltage at time zero. This method of solution is not intuitive to most students and is quite elaborate.

A Systems Thinking approach to this problem would start with a Causal Loop Diagram as shown in figure 21:

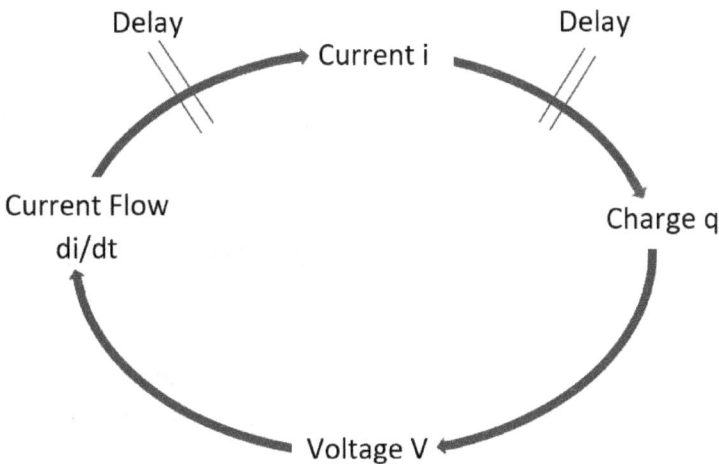

Figure 21. LC Circuit Causal Loop Diagram

The circuit's charge q determines the magnitude of the voltage across both the inductance and the capacitance while simultaneously the voltage impacts the circuit's current flow which impacts the circuit's current which impacts the circuit's charge $q$. However, while the charge instantaneously changes the voltage, the voltage does not

91

instantaneously affect the circuit's charge; there is a delay as the voltage produces a current flow that changes the current and, over time, its charge. It is this delay that causes the oscillation. A stock-and-flow diagram for this situation is shown in Figure 22.

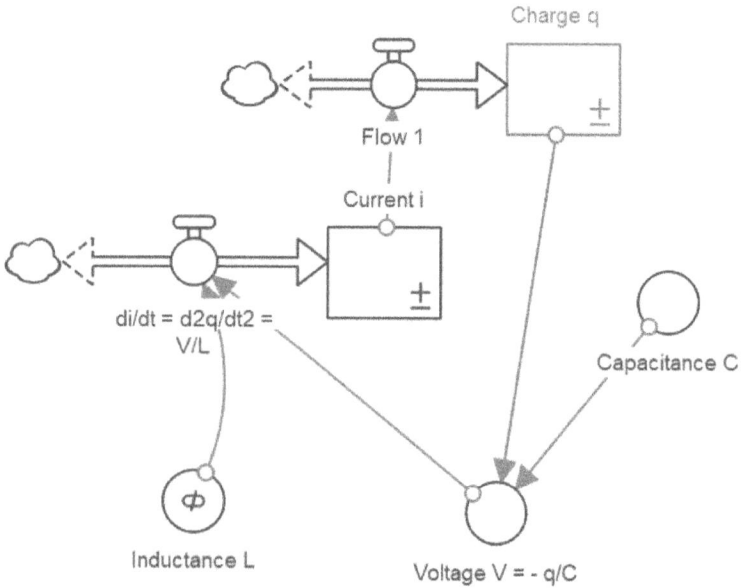

Figure 22. LC Circuit Stock-and-Flow Diagram

The right side of this model shows the voltage generated across the capacitor $V_c = -q/C$ as a result of the charge q. But also $V_c = V_L = L \cdot di/dt$, so that voltage causes a change in the current flow $di/dt$ through the inductor, which is shown on the left side. The voltage responds instantaneously to the charge and so does the $di/dt$, since V=L di/dt. But the current change $di/dt$ causes a change in current i over time (current $i$ does not respond instantaneously to $di/dt$), and the current $i$ causes a charge q change over time (charge q also does not respond instantaneously to current.) Hence current and charge are represented as stocks.

Modeling this system in ISEE Systems' Stella Architect software yields the behavior-over-time plot depicted in Figure 23, showing the sinusoidal oscillation of both current and voltage. No differential equations were involved.

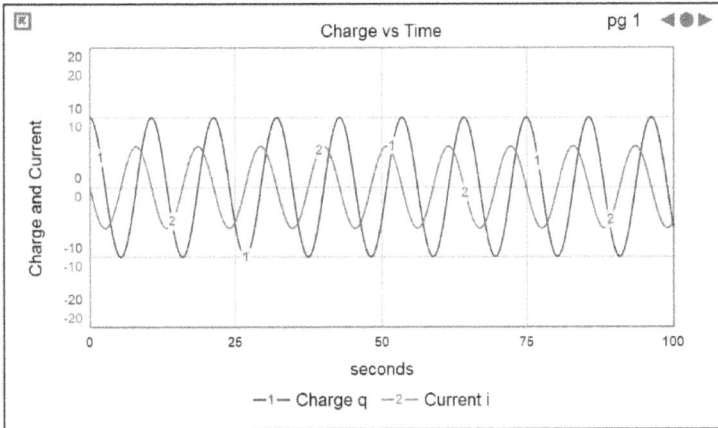

Figure 23. Stella Model Output for LC Circuit

## 4.2.6. Environmental Engineering: Population Dynamics/Epidemiology:
Bacteria Growth. Bacterial growth dynamics are classically described using differential equations.

Definitions:

$N$ = population of bacteria

$N_o$ = initial population

$C$ = carrying capacity in units of population

$t$ = time

$k$ = a rate constant

The rate of change of population N is then written as $dN/dt = kN[1 - N/C]$ which is a non-linear differential equation. This equation is hard to solve using calculus and requires the use of partial fractions. After much effort it eventually leads to the classic logistic solution

$$N(t) = C / \left[1 + \left(\frac{C - N_o}{N_o}\right) e^{-kt}\right] \qquad (9)$$

But System Dynamics provides a much easier and more intuitive solution. A very simple stock-and-flow diagram describing this situation is shown in Figure 24:

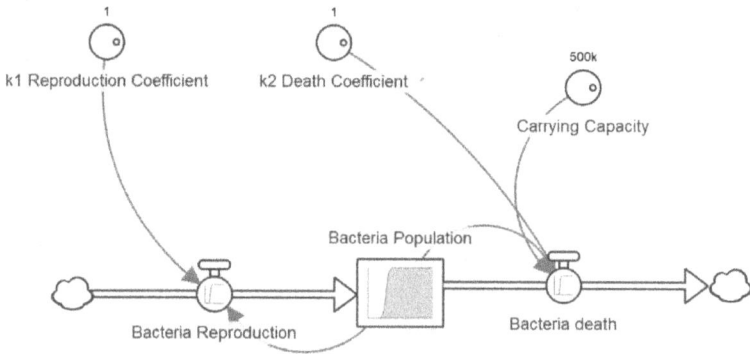

Figure 24. Bacteria Growth Stock-and-Flow Diagram

The bacteria population N is increased by bacteria reproduction and decreased by bacteria death, both of which are impacted by the existing population N. And using the isee Systems Stella Architect software permits the solution of this problem without any differential equations or calculus as depicted in Figure 25:

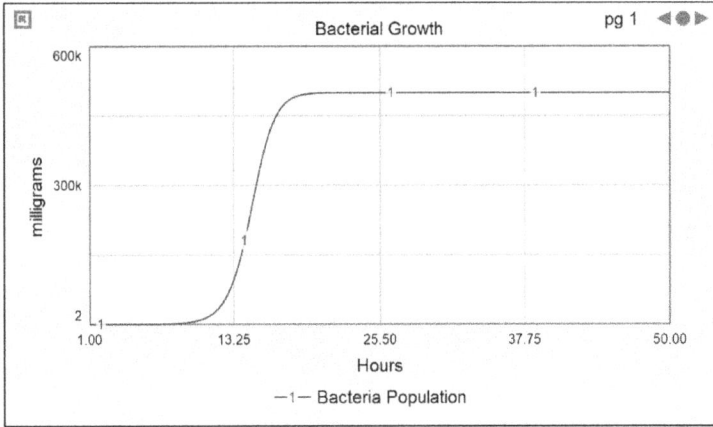

Figure 25. Stella Model Output for Bacteria Growth

*4.2.7. Environmental Engineering: Population Dynamics/Predator-Prey Relationships.* The relationships between predators and prey have been studied extensively by environmental engineers and ecological biologists. The populations of the various species have been described by the Lotka-Volterra differential equations. Where x is the population of prey, y is the population of predators, and b, p, r, and d are constants, these equations are:

$$dx/dt = bx - pxy \qquad (10)$$

$$dy/dt = rxy - dy \qquad (11)$$

These equations *cannot be solved* in closed form; analysists must resort to computer-generated numerical solutions, which typically yield an oscillation in the populations of both predators and prey, with the 2 populations out of synch by a constant amount.

A System Dynamics solution to this problem, on the other hand, is straightforward. We use an example involving cabbages (prey) and rabbits (predators.) The stock and flow diagram is presented in Figure 26.

95

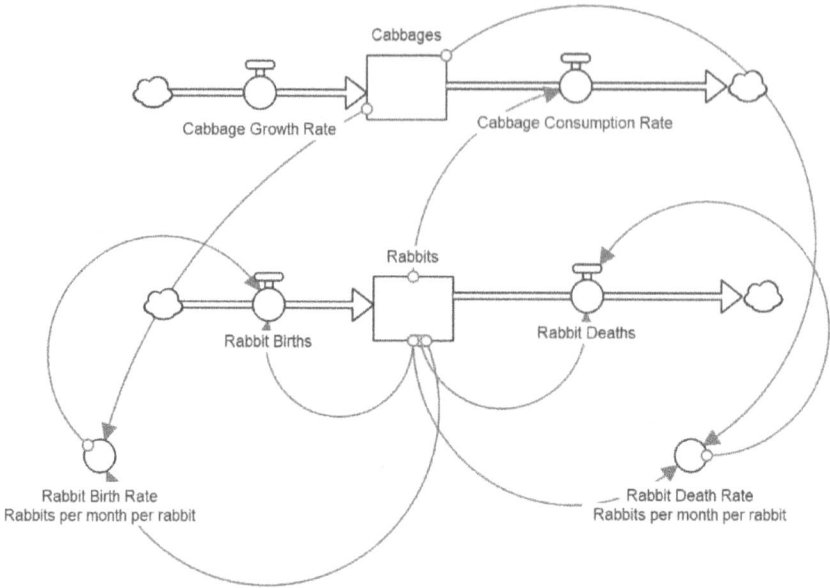

Figure 26. Predator-Prey Stock-and-Flow Diagram

In this model, both cabbages and rabbits are born and die. But the rabbits prey upon the cabbages. When the rabbits have depleted the cabbages, the rabbits decline due to starvation, and the cabbages recover. The new crop of cabbages yields a surge in the rabbit population, and the cycle continues. Modeling this in isee's Stella Architect yields the output shown in Figure 27. The oscillations of the populations of both species are clearly shown.

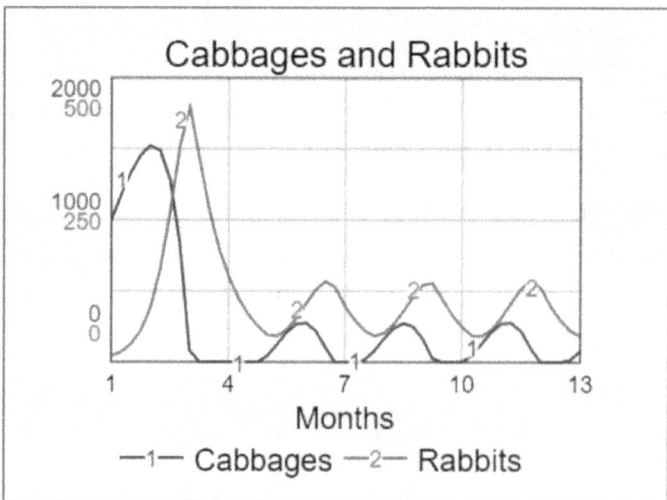

Figure 27. Stella Model Output for Predator-Prey

*4.7.8. Nuclear Engineering: Nuclear chain reaction.* The nuclear fission chain reaction is a reinforcing feedback loop, in which neutrons generated by fission impact atomic nuclei and release more neutrons, which subsequently also impact other nuclei to release even more neutrons. Not all neutrons are active, though: some leak out of the system while some are absorbed by cadmium or boron control rods to control the nuclear reaction. The ratio of Neutron Production Rate to the sum of Neutron Leakage Rate and Neutron Absorption Rate is called $K_{eff}$. If $K_{eff} < 1$ the reaction is sub-critical and the quantity of neutrons decreases exponentially. If $K_{eff} = 1$ the reaction is critical and self-sustaining. If $K_{eff} > 1$ the reaction is super-critical and yields an explosion. A highly simplified stock-and-flow diagram is shown in Figure 28.

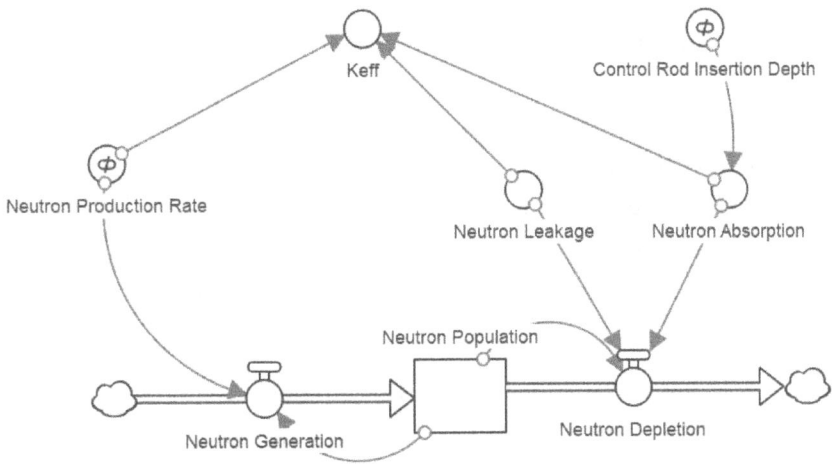

Keff = (Neutron Production rate)/(Neutron Leakage + Neutron Absorption)

Figure 28. Stock-and-Flow Diagram of a Nuclear Reactor

This is a very simplistic model of a nuclear reactor. However, experimenting with the variables is instructive. By adjusting the insertion depth of the control rods, one can cause the reactor to go sub- or super-critical (Figure 29):

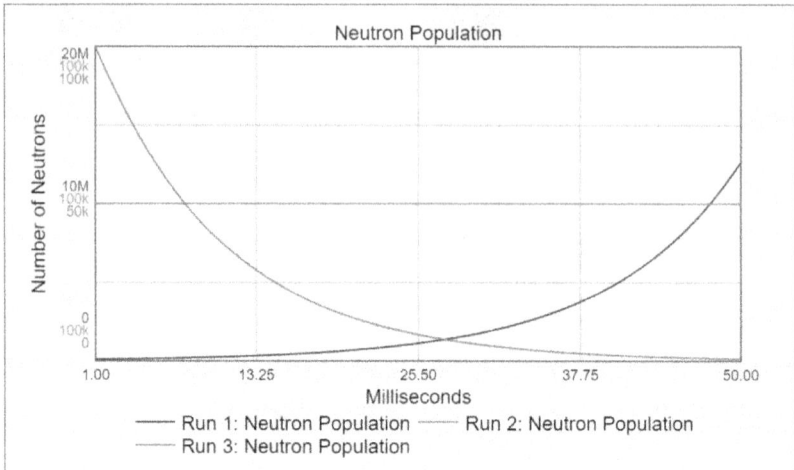

Figure 29. Results of Nuclear Reactor Dynamic Model

Run 1: Control Rod Insertion Depth = 1.7; $K_{eff}$ = 1.05

Run 2: Control Rod Insertion Depth = 1.8; $K_{eff}$ = 1.0

Run 3: Control Rod Insertion Depth = 1.9; $K_{eff}$ = 0.952

In Run 1 with $K_{eff}$ >1 the system is super-critical and the number of neutrons increases exponentially, yielding an explosion. In Run 2 with $K_{eff}$ =1.0, the system is critical, and the number of neutrons is stable. In Run 3 with $K_{eff}$ <1, the system is sub-critical, and the number of neutrons declines exponentially.

In contrast with the simplicity of this model, El-Sefy at al. [35] have constructed a detailed system dynamics simulation of the thermal dynamic processes within different systems in a pressurized water nuclear reactor and have validated the model against other simulations.

For a more in-depth study on applying System Dynamic modeling to typical Engineering science problems, the cited paper [36] and textbook [37] by Hans Fuchs are suggested.

## 4.3. Facilitating Holistic Thinking and Application

Systems Thinking recognizes that technology alone cannot solve complex socio-economic problems. Approaches for addressing world hunger, the need for potable water, transportation systems, the environment, waste management, and many other current issues require more than just technology: solutions to these issues also require consideration of sociological, cultural, economic, philosophical, moral, and political issues: a holistic perspective. An excellent example of the holistic application of Systems Thinking to a real-world potable water issue is provided by *The Water of Ayolé* [38].

Ayolé is a small rural village in the West African country of Togo. The water source for the village in the 1970s-80s was the Amou River, which was infested with the guinea worm *Dracunculus medinensis*, a parasite that infects a human host and causes excruciating pain. To address this issue, government engineers and international aid organizations installed wells in the village. While those wells served the needs of the village for several years, the wells eventually broke down due to normal usage. Unfortunately, no spare parts were available, no technical expertise was available to fix or maintain the pumps, and no money was available to pay for the repairs. As a result, the people of Ayolé were back to using the contaminated water from the river. The government engineers had interpreted this as a purely technical/engineering problem, when in fact this problem was much broader. Fortunately, local Togolese extension agents applied Systems Thinking to address the larger systemic issues. They established a repair parts supply chain via the local Togo hardware store; they trained some of the villagers in well maintenance and repair; and the women of the village developed a farming system that produced and sold agricultural products to generate money for the parts. Several Systems Thinking tools were used to address the Ayolé issue [39].

System Thinking tools that encourage holistic thinking include System Breakdown Structures, System Interrelationship Matrices, and Causal Loop Diagrams. However, a simple awareness of the presence of these non-technical factors resulting in a clearly written paragraph may be just as useful. Many of these issues involve disparate mental models among the various stakeholders. The Iceberg Model (described above) is a useful tool for discovering, exposing, assessing, and revising our ubiquitous (and often incorrect) Mental Models.

*4.3.1. Bounding and Defining the System.* Appropriately defining and bounding the system of interest is vitally important. Several tools are available for this including simple diagrams, tables showing what is in and what is out, or a clearly written paragraph. Had the engineers who installed the well at Ayolé included the villagers and their culture in their system definition, things would likely have proceeded more smoothly.

The System Breakdown Structure (SBS) is a hierarchical pictogram showing the components of a system. It is wise to include in a SBS not only the system proper, but also the environment, users, and support systems that are required to operate and maintain the system; including these exogenous factors results in the depiction of the "suprasystem" and is helpful in identifying the non-technical issues that are likely to impact the system's performance. An example of a SBS applied to *The Water of Ayolé* is shown in Figure 30.

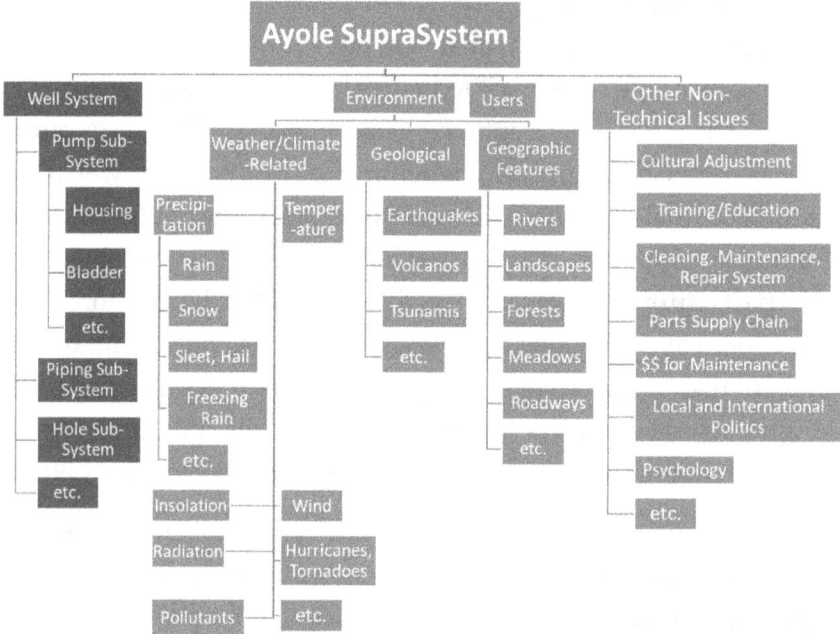

Figure 30. System Breakdown Structure for Ayolé

In Figure 30, the well system itself is shown in red while the exogenous factors are in blue. Had the government agents responsible for the project used a similar SBS, many of the obstacles might have been identified and addressed before problems occurred. The SBS itself is not magical, but if done well it forces engineers to consider non-technical issues affecting the system.

The System Interrelationship Matrix (SIM) is also a pictogram; however, it displays the interrelationships among the system components. A preliminary but incomplete SIM for the highest-level components of *The Water of Ayolé* is shown in Figure 31.

| | Pump | Hole | Mainte nance | Repair Parts | $$ | Village Culture | Educati on | Etc. | Etc. |
|---|---|---|---|---|---|---|---|---|---|
| Pump | - | X | X | X | | X | X | | |
| Hole | X | - | | | X | | | | |
| Mainte nance | X | | - | X | X | X | X | | |
| Repair Parts | X | | X | - | X | | | | |
| $$ | | X | X | X | - | | X | | |
| Village Culture | X | | X | | | - | | | |
| Educati on | X | | X | | X | | - | | |
| Etc. | | | | | | | | - | |
| Etc. | | | | | | | | | - |

Figure 31. System Interrelationship Matrix for Ayolé

The Xs in the figure above indicate where there is a relationship; more detail may be added to each cell to indicate the type of relationship (e.g., electric-mechanical; electric-chemical; mechanical-psychological.) In addition, SIMs may be developed for lower levels of the system to show greater detail at sub-levels. A simple SIM like this would have demonstrated (for example) the interactions between the well maintenance system and the village culture.

Additional tools such as the Causal Loop Diagram may be applied to capture different perspectives. Figure 32 shows Causal Loop Diagrams describing two perspectives on the Ayolé scenario:

Water of Ayolé Causal Loop Diagram: Initial

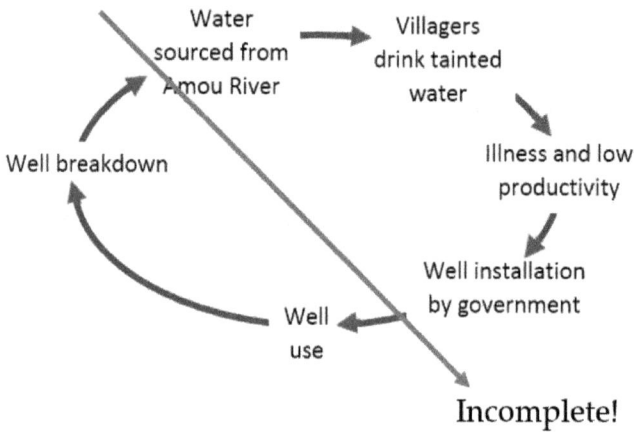

Water of Ayolé Causal Loop Diagram: Final

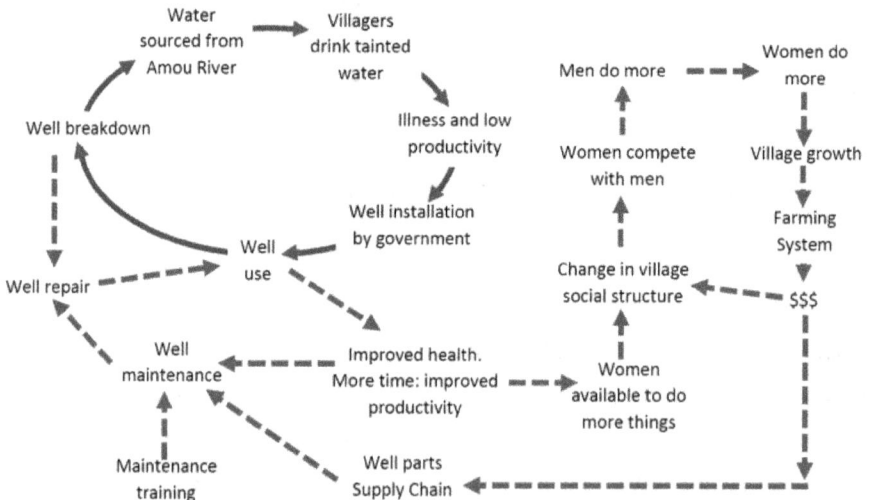

Figure 32. Causal Loop Diagrams for the Water of Ayolé

What was thought to be a simple engineering problem was really an engineering/socio-economic/logistics/ psychological problem, as illustrated in the Causal Loop Diagrams above in Figure 32 [Monat & Gannon, 2018].

The Iceberg Model has been described above. It is especially useful for discovering, exposing, assessing, and revising mental models. Figure 33 shows an Iceberg Model for Ayolé from the engineers' perspective.

Figure 33. Ayolé Iceberg Model—Engineers' Perspective

Figure 34, on the other hand, shows the corresponding Iceberg Model from the villagers' perspective.

Figure 34. Ayolé Iceberg Model—Villagers' Perspective

Clearly, the disparate mental models conflict and must be resolved. Other Iceberg Models from the perspectives of politicians, the U.S. A.I.D., and others are also relevant and should be included.

One cannot say with certainty that the application of these Systems Thinking tools would have prevented the Water of Ayolé problems. However, they would have facilitated the consideration of non-technical issues and would have exposed the confounding mental models, thus increasing the probability of success. Engineers should be taught these tools to facilitate holistic thinking.

## 5. Preliminary Assessment/Field Study.

We have not performed a controlled experiment to validate our hypotheses, nor have we yet conducted a case study or attempted to adapt the curriculum at Worcester Polytechnic Institute per our recommendations. However, we have developed and taught an undergraduate Systems Thinking course 3 times, and the feedback from those offerings is supportive. ES1500 *Introduction to Systems Thinking* was taught during the early springs of 2019, 2020, and 2021; Table I shows student evaluations.

Table I. Results of First 3 Offerings of *Introduction to Systems Thinking*

| Term | Enrollment | Average Student Rating of the Overall Course Quality (out of 5.0) |
|------|------------|------------------------------------------------------------------|
| 2019 | 20 | 5.0 |
| 2020 | 20 | 4.8 |
| 2021 (Remote) | 24 | 4.6 |

Student comments include, "Fascinating course, should be required for most majors, gives you important insights about how many things in the world work, and how to avoid pitfalls in products/designs you make," "It's a fascinating subject and one I think can apply to many fields, not just civil engineering," "Systems Thinking is an extremely exciting field that is often overlooked at WPI. It was great to have a course that studies this topic that is relevant in all our lives," "I learned so much taking this course and now I feel like I look at the world around me with a systems thinking perspective, which I didn't have before," and "New class was handled well and the subject matter was made relevant to all majors." These results are not a statistical validation of our proposals; however they represent an initial assessment of the perceived value of Systems Thinking to undergraduates. Future plans include a quantitative assessment of our positions.

## 6. Proposal for Integrating Systems Thinking into Engineering Education

Many administrators attempt to include Systems Thinking in engineering education by simply adding a Systems Thinking course or two to their undergraduate engineering curriculum. In our opinion, this is a necessary but not sufficient approach. Although undergraduate engineering curricula should certainly include Systems Thinking courses, instructors need to demonstrate the application of Systems Thinking to engineering problems in traditional engineering courses.

Like calculus, Systems Thinking is a perspective, a language, and a set of tools that can and should be applied to most engineering problems. And like calculus, it should be used whenever engineering disciplines are taught, not as a replacement for traditional engineering approaches, but as an adjunct to them.

In their regular courses on Mechanical, Chemical, Electrical, Civil, Environmental, and Aerospace Engineering, instructors must seek opportunities to demonstrate the application of Systems Thinking. Systems Thinking is applicable in the following situations:

a. Whenever a differential equation is used, a System Dynamics model may be used in addition.

b. Whenever a solution is sought to an infrastructure or design problem, Systems Thinking should be applied, as complex socio-economic problems cannot be solved by technology alone.

To overcome resistance to inculcating Systems Thinking into undergraduate engineering education, we recommend the following:

a. Engineering instructors and administrators should educate themselves about Systems Thinking tools and their application. Several courses are available, as are several good books on the subject. We recommend *Thinking in Systems* by Dana Meadows, *Systems Thinking* by Daniel Kim, *An Introduction to Systems Thinking with iThink* by Barry Richmond, and *Using Systems Thinking to Solve Real-world Problems*, by Jamie Monat and Thomas Gannon.

b. College administrators should invoke the teaching of Systems Thinking as a means to support ABET standards.

c. Systems Thinking researchers and practitioners should continue to demonstrate and publish the beneficial results accruing from the application of Systems Thinking to engineering problems. Initial applications can focus on incorporating conceptual modeling approaches such as, Causal Loop, and Stock and Flow diagrams into typical engineering texts, such as [40,41] that address dynamic modeling of systems.

d. College administrators should require that Systems Thinking be taught as a part of every engineering program---not as a stand-alone course or courses, but integrated with the other engineering disciplines, just as calculus and physics are.

e. K-12 Teachers should introduce Systems Thinking concepts to their students so that the expectation is set to learn Systems Thinking in College, which would drive the demand for including Systems Thinking.

f. Industry advisory boards to university engineering departments should stress the importance of Systems Thinking concepts in undergraduate engineering education.

## 7. Conclusions/Limitations

The role of engineers is changing. No longer can they make engineering decisions independent of psychological, sociological, and environmental impacts. The interconnectedness of all things and people mandates that engineers take a holistic view, understand the second and third-order impacts of their designs, and recognize that there is usually not a "best" design but instead several acceptable designs involving various trade-offs. Traditional engineering educational methods involving algebraic plug-and-chug formulas, static models, and differential equations do not provide this holistic view or the insights and sensitivity required by today's engineers.

Systems Thinking, on the other hand, addresses these limitations. It provides a holistic view of engineered systems and explores the relationships among system components, users, support infrastructure, and the environment. It uses dynamic models to understand the behavior of systems over time, causal loop and stock-and-flow diagrams to understand interrelationships, and the Iceberg Model to understand the underlying forces, structures, and mental models that drive system behavior. It emphasizes trade-offs and stochastic solutions as opposed to deterministic black-and-white solutions while providing insight and understanding not available from traditional engineering approaches.

But there are obstacles to the inculcation of Systems Thinking into undergraduate engineering education: its methodology, tools, and benefits are not well-understood by instructors; it is perceived to be a substitute or replacement for conventional, tried-and-true engineering approaches; and engineering curricula are often full and cannot accommodate additional courses. To overcome these, we suggest that

Engineering instructors and administrators educate themselves about Systems Thinking tools and their application; that Systems Thinking researchers and practitioners continue to demonstrate and publish the beneficial results accruing from the application of Systems Thinking to engineering problems; and that college administrators require that Systems Thinking be taught as a part of every engineering program---not as a stand-alone course or courses, but integrated with the other engineering disciplines. In this paper, we detail methods and examples that should facilitate this.

## References

1. Baldessari, R.; Bödekker, B.; Deegener, M.; Festag, A.; Franz, W.; Kellum, C.C.; Kosch, T.; Kovacs, A.; Lenardi, M.; Menig, C. Car-2-car communication consortium-manifesto. **2007**.

2. De Silva, L.C.; Morikawa, C.; Petra, I.M. State of the art of smart homes. *Engineering Applications of Artificial Intelligence* **2012**, *25*, 1313-1321.

3. Monat, J.P. The emergence of humanity's self-awareness. *Futures* **2017**, *86*, 27-35.

4. ABET. Criteria for Accrediting Engineering Programs, 2019 – 2020. Available online: https://www.abet.org/accreditation/accreditation-criteria/criteria-for-accrediting-engineering-programs-2019-2020/#GC2 (accessed on May 1, 2021).

5. Hadgraft, R.G.; Carew, A.L.; Therese, S.A.; Blundell, D.L. Teaching and assessing systems thinking in engineering. In Proceedings of Research in Engineering Education Symposium; pp. 230-235.

6. Davidz, H.L.; Nightingale, D.J. Enabling systems thinking to accelerate the development of senior systems engineers. *Systems Engineering* **2008**, *11*, 1-14.

7. Muci-Kuchler, K.H.; Bedillion, M.; Huang, S.; Degen, C.; Ellingsen, M.; Nikshi, W.; Ziadat, J. Incorporating basic systems thinking and systems engineering concepts in a mechanical engineering sophomore

design course. In Proceedings of ASEE Annual Conference, Columbus, Ohio **2017**.

8. Valenti, M. Teaching tomorrow's engineers. *Mechanical Engineering* **1996**, *118*, 64.

9. Employment and Training Administration, US Department of Labor. Engineering Competency Model. Available online: https://www.careeronestop.org/CompetencyModel/competency-models/pyramid-download.aspx?industry=engineering (accessed on May 1, 2021)

10. Whitcomb, C.A.; Delgado, J.; Khan, R.; Alexander, J.; White, C.; Grambow, D.; Walter, P. *The Department of the Navy systems engineering career competency model*; Naval Postgraduate School Monterey CA, Graduate School of Business and Public Policy. 2015.

11. Crawley, E.F.; Malmqvist, J.; Lucas, W.A.; Brodeur, D.R. The CDIO syllabus v2. 0: An updated statement of goals for engineering education. In Proceedings of Proceedings of 7th international CDIO conference, Copenhagen, Denmark. 2011.

12. Yurtseven, M.K. Teaching Systems Thinking to Industrial Engineering Students. In Proceedings of Ist International Conference on Systems Thinking in Management. 2000.

13. Rehmann, C.R.; Rover, D.T.; Laingen, M.; Mickelson, S.K.; Brumm, T.J. Introducing systems thinking to the Engineer of 2020. Iowa State University, Digital Repository, Agricultural and Biosystems Engineering Conference Proceeding and Presentations. **2011**.

14. Degen, C.M.; Muci-Küchler, K.H.; Bedillion, M.D.; Huang, S.; Ellingsen, M. Measuring the Impact of a New Mechanical Engineering Sophomore Design Course on Students' Systems Thinking Skills. In Proceedings of ASME 2018 International Mechanical Engineering Congress and Exposition.

15. Robinson-Bryant, F. Developing a Systems Thinking Integration Approach for Robust Learning in Undergraduate Engineering Courses. *American Society for Engineering Education.* **2018**.

16. Cattano, C.; Nikou, T.; Klotz, L. Teaching systems thinking and biomimicry to civil engineering students. *Journal of Professional Issues in Engineering Education & Practice* **2011**, *137*, 176-182.

17. Hung, W. Enhancing systems-thinking skills with modelling. *British Journal of Educational Technology* **2008**, *39*, 1099-1120.

18. Jain, R.; Sheppard, K.; McGrath, E.; Gallois, B. Promoting Systems Thinking in Engineering and Preengineering Students. in *ASEE Annual Conference and Exposition,* 2009.

19. Wasson, C.S. System Engineering Competency: The Missing Element in Engineering Education. In Proceedings of Proceedings of the 20th Anniversary of the INCOSE International Symposium Chicago, IL (US). 2010.

20. Summerton, L.; Clark, J.H.; Hurst, G.A.; Ball, P.D.; Rylott, E.L.; Carslaw, N.; Creasey, J.; Murray, J.; Whitford, J.; Dobson, B. Industry-informed workshops to develop graduate skill sets in the circular economy using systems thinking. *Journal of chemical education* **2019**, *96*, 2959-2967.

21. Voorhees, K.; Hutchison, J.E. Green Chemistry Education Roadmap charts the path ahead. Amer Chemical Soc, 1155 16th St, NW, Washington, DC 20036 USA: 2015.

22. Michael, B. What Is the Relationship Between Systems Thinking and Lean? Available online: https://thesystemsthinker.com/%EF%BB%BFwhat-is-the-relationship-between-systems-thinking-and-lean/ (accessed on May 19, 2021).

23. Monat, J.P.; Gannon, T.F. What is systems thinking? A review of selected literature plus recommendations. *American Journal of Systems Science* **2015**, *4*, 11-26.

24. SEBoK Editorial Board; Cloutier, R.J. (Editor in Chief). The Guide to the Systems Engineering Body of Knowledge (SEBoK), v. 2.4. Available online: (accessed on May 10, 2021).

25. Robinson, S. Conceptual modelling for simulation Part I: definition and requirements. *Journal of the operational research society* **2008**, *59*, 278-290.

26. Forrester, J.W. Lessons from system dynamics modeling. *System Dynamics Review* **1987**, *3*, 136-149.

27. Forrester, J.W. System dynamics: the foundation under systems thinking. *Sloan School of Management. Massachusetts Institute of Technology* **1999**, *10*.

28. Alexandra, B.S.; Peggy, O.; Roseanne, S. Preventing Culture from Eating Your Strategy. In *Global Impact Collaboratory: Monitoring, Evaluation, and Learning*, 2020; Vol. 2021.

29. Monat, J.P.; Gannon, T.F. *Using Systems Thinking to Solve Real-World Problems*; College Publications: 2017.

30. Meadows, D.H. *Thinking in systems: A primer*; chelsea green publishing: 2008.

31. Forrester, J.W. Industrial dynamics. *Journal of the Operational Research Society* **1997**, *48*, 1037-1041.

32. Forrester, J.W. Urban dynamics. *IMR; Industrial Management Review (pre-1986)* **1970**, *11*, 67.

33. Forrester, J.W. *World dynamics*; Wright-Allen Press: 1971.

34. Richmond, B.; Peterson, S. *An introduction to systems thinking*; High Performance Systems., Incorporated Lebanon, NH: 2001.

35. El-Sefy, M.; Ezzeldin, M.; El-Dakhakhni, W.; Wiebe, L.; Nagasaki, S. System dynamics simulation of the thermal dynamic processes in nuclear power plants. *Nuclear Engineering and Technology* **2019**, *51*, 1540-1553.

36. Fuchs, H.U. System dynamics modeling in science and engineering. *Centro de investigación para la ciencia e ingeniería, Mayaguez: Sistemas dinámicos en la Universidad de Puerto Rico* **2006**.

37. Fuchs, H.U. *The dynamics of heat: a unified approach to thermodynamics and heat transfer*; Springer Science & Business Media: 2010.

38. Nichols, S.; Klein, L. The Water of Ayolé : Togo, West Africa. Iowa State University: Ames, Iowa, 2008.

39. Senge, P.M.; Kleiner, A.; Roberts, C.; Ross, R.; Smith, B.J. The Fifth Discipline Fieldbook (New York, Currency Doubleday). **1994**, 156-161.

40. Lobontiu, N. *System dynamics for engineering students: Concepts and applications*; Academic Press: 2017.

41. Palm, W.J. *System dynamics*; McGraw-Hill: 2010; Vol. 2.

# Practical Applications of Systems Thinking to Business

Jamie Monat *, Matthew Amissah and Thomas Gannon

Electrical & Computer Engineering Department (Systems Engineering Program), Worcester Polytechnic Institute, Worcester, MA 01609, USA;

*reprinted from Systems 2020, 8, 14.*

**Abstract:** In this paper we summarize the research on Systems Thinking for business management and explore several examples of business failures due to a lack of application of Systems Thinking, with an ultimate goal of offering a Systems Thinking approach that is useful to all levels of management. Although there is significant literature aimed at facilitating Systems Thinking in organizational management, there remains a lack of adoption of Systems Thinking in mainstream business practice. This is perhaps because the literature does not reduce high-level Systems Thinking principles to hands-on, practical protocols that are accessible for typical managers, thus limiting the working application of Systems Thinking concepts to researchers and consultants who specialize in the field. The goal of this work is to not only elaborate on the high-level ideals of System Thinking, but also to articulate a more precise and practical hands-on approach that is useful to all levels of business managers.

**Keywords:** systems thinking; business; complexity

## 1. Introduction

Systems Thinking has been characterized as a perspective, a language, and a set of tools that can be useful in solving complex problems that are typically not amenable to conventional reductionist thinking. Applications have been proposed in engineering, politics, international relations, biology, astrophysics, economics, etc. In this paper, we focus on the application of Systems Thinking to business. We summarize the research on Systems Thinking for organizational management, explore several examples of business failures due to a

113

lack of application of Systems Thinking principles, and conclude with a practical, hands-on protocol that facilitates the application of Systems Thinking to business decisions. In light of the increasingly complex climate within which business managers operate, this paper highlights the need for and benefits of a practical discipline of Systems Thinking in business.

There is a significant body of literature (spanning a century of research) aimed at facilitating Systems Thinking in organizational management, yet there remains a lack of adoption of these approaches in mainstream business practice. We believe that this is because the high-level principles of Systems Thinking in organizations have not been reduced to hands-on rules and policies that are actionable by business managers. Although concepts such as the application of cybernetics to organizational viability, the benefits of dynamic modelling to understand systemic structure, the importance of shareholder participation in decision-making, and the importance of a holistic perspective are certainly valid, business managers need to know how Systems Thinking informs pricing decisions, product development, staffing decisions, sales commission structures, inventory management, quality assurance procedures, administrative decisions, and employee motivation. The goal of this work is to not only elaborate on the high-level ideals of System Thinking that offer a useful bird's eye perspective, but also to articulate a practical approach that is useful to all levels of business managers.

We adopt an inductive approach to the subject. Whereas the more common deductive approach would start with a theoretical framework for business and then develop and test a hypothesis such as "the application of Systems Thinking principles results in better business success", we do not believe that we can make a convincing case using this reasoning. Instead, we start with business observations and then, using a Systems Thinking lens, work backwards toward generalizations and principles that are useful for business practitioners.

Following this inductive approach, we first identify specific failure or problem examples organized based on principal areas common to many businesses, namely: 1. product development and product lifecycle management, 2. sales and sales management, 3. pricing, 4. operations and quality assurance, and 5. administration. Where possible, we use famous examples that have been widely publicized. Where we cannot find famous examples, we tap our personal observations covering over 70 combined years of industrial business management experience and identify real examples from our own experience. We conclude each example with a discussion of Systems Thinking lessons that could have been implemented to avert the failure. Following the examples, we identify commonalities and ultimately synthesize these into a protocol that can be readily applied by decision-makers in similar contexts.

Although the approach adopted is mostly informal, it offers the benefit of effectively communicating lessons in Systems Thinking using well known business failures. The goal is to elevate Systems Thinking concepts to the level of a business management *paradigm* for the practitioner. This, we hope, shall offer managers a protocol for business that exploits the benefits of Systems Thinking, minimizes common errors, and maximizes profits.

Subsequent sections of the paper are organized as follows: Section 2 offers definitions and a review of Systems Thinking concepts. Section 3 entails a brief discussion of the literature on Systems Thinking in the context of organizations. Section 4 discusses examples of business failures that could have been averted with Systems Thinking. Section 5 discusses a practical, hands-on approach to implementing Systems Thinking in business management. In Section 6 we offer a summary of the paper and comment on expected contributions.

## 2. What is Systems Thinking?

Systems Thinking (ST) is a perspective, a language, and a set of tools [1]. It is a holistic perspective that acknowledges that the relationships among system components and between the components and the environment are as important (in terms of system behavior) as the components themselves. It is a language of feedback loops, emergent properties, complexity, hierarchies, self-organization, dynamics, and unintended consequences. Systems Thinking tools include the Iceberg

Model [2,3], which posits that in human-designed systems, repeated events and patterns (which are observable) are caused by systemic structures (stocks, flows, and feedback loops) which are, in turn, caused by underlying mental models, which are often hidden. Examples of systemic structures include organizational hierarchy; social hierarchy; interrelationships; rules and procedures; authorities and approval levels; process flows and routes; incentives, compensation, goals, and metrics; and corporate culture. Behaviors are derived from these structures, which are (in turn) established due to mental models. A related fundamental Systems Thinking concept is that different people in the same structure will produce similar results. In order to understand behaviors, one must first identify and then understand the systemic structures and underlying mental models that cause them. The Iceberg Model is illustrated in Figure 1.

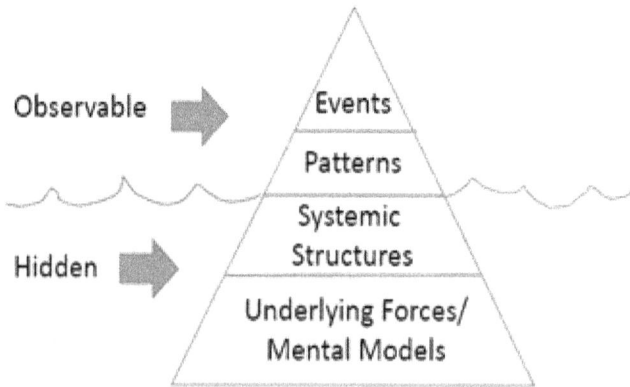

**Figure 1.** The Iceberg Model [1].

Additional Systems Thinking tools include causal loop diagrams, behavior-over-time plots, stock-and-flow diagrams, systemic root cause analysis (which often leads to culture as a systemic root cause), dynamic modeling tools, and archetypes. A more comprehensive explanation of Systems Thinking is provided by [1].

According to Monat [4], a system is a group of interacting, interrelated, or interdependent parts that together form a unified whole, for which the arrangement of the parts is significant, that has constraints and boundaries, and that attempts to maintain stability through feedback. Clearly, a business is a system. Yet, few people acknowledge this or use Systems Thinking to manage their businesses.

Inasmuch as businesses are rife with feedback loops, unintended consequences of business decisions, mental models and concomitant structures, and oscillations of inventory and staff levels, they are well-suited to Systems Thinking tools and concepts. Systems Thinking offers a lens for establishing and examining purpose, structure, and behavior in the context of an increasingly complex dynamic environment. In the following two sections we briefly review the literature and discuss some common business failures that occur as a result of poor Systems Thinking.

## 3. Literature Review

There is a significant body of research on the application of Systems Thinking in the context of managing business organizations. For the purpose of this work we offer an exploratory review that covers the main research paradigms and notable methodologies proposed in the literature. An outline of the literature is given here from the perspective of three research paradigms adapted from the work of Checkland [5] and Jackson [6], namely: Hard Systems Thinking (HST), Soft Systems Thinking (SST), and Critical Systems Thinking (CST). It is important to note that these paradigms are employed here to offer a cursory retrospection of the evolution of Systems Thinking (from 1930–1960s, 1970s, and 1980s, respectively) and are not intended to delineate hard categories applicable to all of the research literature.

The underlying assumption in HST is that human organizations can be objectively studied, modeled, and controlled to meet defined purposes. Thus, HST approaches are mainly driven by the development of models to understand organizations and inform on optimal courses of action to achieve end goals. Checkland [7] alluded to limitations in the treatment of ill-structured problems using HST approaches then espoused in disciplines such as Cybernetics, General Systems Theory, Systems Engineering, and Operations Research. An example HST approach is Viable Systems Diagnosis based on Stafford Beer's work in Management Cybernetics [8-10]. Beer's work sought to explain how systems are capable of maintaining their independent existence [11]. His Viable System Model (VSM) applies Cybernetics principles to specify, essentially, the pre-conditions of organizational viability. The VSM has been applied extensively as a tool for diagnosis and design of various business organizations. A review of the VSM and several case studies applying the model to organizational problems is given in [12,13].

Another HST approach initially developed in the late 1950s is Systems Dynamics. Jay Forrester's seminal work [14] extended the mathematics of feedback control systems to support simulation-based analysis for managerial decision-making. Later works in [15] and [16] extended the approach from the corporate world to modeling a broader context of social systems in order to inform urban and global policy decisions. Systems Dynamic models have been applied to various business contexts. A classic example is the Beer Game [17], which, along with several others [18], is commonly used as a training tool demonstrating that business performance is often dictated by an underlying structure as opposed to random external factors or events that are out of the control of decision-makers. A compilation of case studies applying System Dynamic models to real world problems is offered by the System Dynamics Society [19]. Other HST approaches include applications of Complexity Theory to business [20,21], Agent-Based Modeling, and Simulation [22]. See Table 1 for a summary list of references for HST approaches.

SST, largely based on the work of Checkland, entails the fundamental assumption that managerial problems are typically ill-structured with several possible interpretations based on the perspective taken by an observer. SST thus emphasizes that the process of inquiry into such problem situations should be organized as a learning system that integrates divergent stakeholder perspectives. An example SST approach is Checkland's Soft Systems Methodology (SSM) [23]. It proposes an iterative process that develops and applies models under various stakeholder viewpoints to facilitate learning about a problem context in order to support decision-making. A more expansive review including case studies is provided in [24,25]. We identify other Systems Thinking methodologies that emphasize stakeholder participation and facilitating shared agreement of the problem situation under the SST paradigm. Examples include Interactive Planning [26] and Team Syntegrity [27]. The complete list of approaches identified is summarized in Table 1.

Finally, CST, embodied by the work of Jackson and Flood [6,28], argues for embracing multiple ST methodologies and selectively applying them based on the problem context. CST advocates assessing the problem situation and identifying inherent viewpoints, assumptions, biases, etc., while also understanding the available methodologies, their strengths, weaknesses, and implications for adoption. The Total Systems Intervention (TSI) methodology presents an approach for CST practice. It entails iterative phases: Creativity, Choice, and Implementation for problem characterization, selection of applicable methodologies, and implementation, respectively. An extensive review of TSI with example case study applications is offered in [29]. Table 1 presents an overview of the Systems Thinking paradigms discussed and examples of methodologies aligned with them.

**Table 1.** Overview of Business-related Systems Thinking Paradigms.

| Paradigm | Description | Methodologies |
|---|---|---|
| **Hard Systems Thinking** | Systems Thinking (ST) essentially applies various models to understand and ultimately control the structure and consequent behavioral patterns constituting an identified problem situation | General Systems Theory [30,31]<br><br>Viable System Diagnosis [8-10]<br><br>Systems Dynamics [15,16,32]<br><br>Enterprise Systems Engineering [33]<br><br>Complexity Theory [20,21]<br><br>Agent-Based Modeling and Simulation [22] |
| **Soft Systems Thinking** | ST essentially captures stakeholders' viewpoints, facilitates dialogue, learning and plans for implementing interventions in a problem situation. | Soft Systems Methodology [23]<br><br>Interactive Planning [26]<br><br>Social Systems Design [34,35]<br><br>Strategic Assumption Surfacing and Testing [36] |

| | | Team Syntegrity [27] |
| --- | --- | --- |
| | | Cognitive Mapping [37] |
| | | Hypergame Analysis [38] |
| **Critical Systems Thinking** | ST entails critically evaluating a problem situation, divergent stakeholder worldviews of the situation while also assessing the strengths and weaknesses of candidate methodologies suitable for the problem context | Total Systems Intervention [29]<br><br>Critical Heuristics [39] |

As expected, more recent works in the field are integrative, supporting more than one paradigm. These are not only aimed at making recommendations for business decisions using models, but also exposing and shaping the mental models of stakeholders and guiding the overall problem-solving process in organizations. A classic example is Peter Senge's *The Fifth Discipline* [40], which arguably sparked a resurgence in attention and adoption of Systems Thinking-motivated practices in business organizations. Senge defines Systems Thinking as a discipline for seeing wholes, a framework for seeing interrelationships rather than things, and for seeing patterns of change rather than snapshots. The perspective of Systems Thinking taken in the book is mostly based on System Dynamics, however it is not presented as a stand-alone approach. Systems Thinking is presented as a key discipline that integrates four others: Personal Mastery, Mental models, Building Shared Vision, and Team Learning. Senge identifies these disciplines as necessary for a learning organization.

A follow up book, *The Fifth Discipline Fieldbook* [2], offers a much more applied exploration of ST in business. It presents many practical examples and case studies, along with a problem-solving methodology. Similarly, Ballé [41] aimed at a practical guide to Systems Thinking in business, rooted in Systems Dynamics and Senge's work. He observed that a typical management reaction to an issue is a myopic, short-term solution instead of a long-term systemic analysis. This usually results in a pattern of recurrent problems that are supposed to have been solved in the past. Systems Thinking is proposed as the key to intuitive and sustainable solutions for addressing everyday business problems.

It is only proper that we end this review with a reference to Norman's article [42] on Systems Thinking in product development. The article proposes that products should be thought of as not just the physical entity, but rather as a service that encompasses the experience of researching, shopping, buying, using, and maintaining the product. Several products are given as examples to reinforce this idea: the Mini-Cooper, Amazon's Kindle, the iPod, etc. For example, he attributes the iPod's success not only to the physical device's aesthetics and functionality, but also to the music downloading service, digital rights management system, Apple's Genius Bar, etc., all of which combine to provide an excellent user service experience.

While the literature shows a sustained research effort spanning a century of methodologies and applications to enhance Systems Thinking, the adoption of these methodologies in the mainstream of business practice is lacking. In the subsequent sections we review several examples of major business failures which could have been avoided with Systems Thinking. This highlights the challenge of readily accessible Systems Thinking constructs that can be implemented as a guidance protocol to support the day-to-day decision-making involved in running a business.

## 4. Business and Business Component Failures and Systems Thinking Lessons

In this section we discuss several business failures, most of which are widely known, either because they have been well publicized in the media or are fairly common scenarios in business. The examples provided are drawn from the following principal areas: product development and product lifecycle management, sales and sales management, pricing, operations and quality assurance, and administration.

Each example concludes with a discussion of Systems Thinking lessons that could have been implemented as a strategy to avoid the failure. We must point out that the identified Systems Thinking lessons are not meant to imply that these problems have a simple or unique cause or fix. Rather, these observations underlie the broader theme that problems are often caused and/or reinforced by underlying organizational structures and the mental models of stakeholders involved. Therefore, decision-makers' responses to such problems should be informed by this broader perspective.

*4.1. Examples in Product Development and Product Lifecycle Management*

4.1.1. The Microsoft Zune

Apple released its spectacularly successful iPod in 2001. By 2005, sales exceeded 20 million units per year. To compete with Apple, Microsoft released its Zune personal music player in 2006 (shown in Figure 2). However, the Zune did not share the iPod's ergonomics or aesthetics. Moreover, Microsoft had not developed all the ancillary systemic structures that made the iPod successful.

**Figure 2.** Apple's iPod and Microsoft's Zune.

For the iPod to really catch on, a system had to be developed that facilitated the downloading of music from the Internet. This required not only the technology, but also consideration of ancillary factors, such as licensing, royalties, payment and transaction management, and storage. By addressing each of these with the development of

122

iTunes, Apple not only enabled the iPod, but completely disrupted existing music listening technology (CDs).

Why is this Systems Thinking? In Designing for People, Don Norman says, "It is not about the iPod; it is about the system. Apple was the first company to license music for downloading. It provides a simple, easy to understand pricing scheme. It has a first-class website that is not only easy to use but fun as well. The purchase, downloading the song to the computer and thence to the iPod are all handled well and effortlessly. And the iPod is indeed well designed, well thought out, a pleasure to look at, to touch and hold, and to use. There are other excellent music players. No one seems to understand the systems thinking that has made Apple so successful."[42]

The iPod is not a stand-alone product; instead, it is part of a personal entertainment system, the elements of which include the iPod itself, the individual who is listening, the environment (indoors, outdoors, office, gym, etc.), the songs, the song acquisition and storage, and the activities while listening (whether jogging, studying, relaxing, spinning, driving, etc.). The iPod Personal Entertainment System is not a product at all; it is a service. It is experienced, not consumed. Apple's recognition of this and that the device itself is simply an element of this service was not only innovative, but revolutionary. While other manufacturers (Sony, Tascam, Microsoft, Diamond, etc.) structured their companies to support the device that they manufactured, Apple structured their company to support the user.

*Systems Thinking Lesson*: Failure to understand that many products are not really stand-alone devices, but instead are merely one component of a user experience system.

### 4.1.2. Polaroid

Polaroid rose to market dominance in instant photography as a result of the innovative products developed by its founder, Edwin Land, and was once considered a stellar example of a high-tech success [43]. In the 1980s, Polaroid was well aware of the emerging trend in electronic imaging technology, and by 1989 was spending 42% of its research and development budget on digital imaging. Polaroid was the number one provider of digital cameras by 1990.

Despite its early lead in the digital camera market, Polaroid failed to take full advantage of the emerging trends in digital photography, such as purely digital workflow. Senior management continued to rely on an outdated mental model that customers wanted hard-copy print, rather than electronic images that could be viewed on digital displays and in slide shows. Even though Polaroid made significant investments in its Micro-Electronics Laboratory in the mid-1980s, the company had a bias against electronics. That bias was fueled by the significant profitability of their film business with gross margins of over 65%, which made the consideration of new business models and markets out of the question (a frozen paradigm). Over time, customers began to realize the speed and cost-savings associated with digital workflow, as digital cameras became commodities and resolution increased. As a result, Polaroid began losing its largest customers in the real estate and photo-identification markets, and its sales of film dropped precipitously. By October 2001, Polaroid filed for bankruptcy and never recovered.

*Systems Thinking Lesson:* This is an example of a flawed mental model. Although Polaroid instructed its researchers to develop digital cameras in response to emerging trends in digital photography, the focus was on developing digital cameras that could produce hard-copy print, without recognizing emerging new market opportunities. Polaroid's innovation was limited by its out-of-date mental models/frozen paradigm, while competitors such as Canon, Nikon, and even Kodak (to a degree) were better Systems Thinkers.

4.1.3. Research in Motion's Blackberry

The Blackberry smart phone (see Figure 3) was launched by Canadian company Research in Motion (RIM) in 1999 [44].

**Figure 3.** The Blackberry.

RIM had the lion's share of the smartphone market by 2007, with over 10 million subscribers, and was worth over $67 Billion. The success of the Blackberry smart phone was due to several innovative features, such as the Blackberry Messenger (BBM) service and Blackberry Curve, strong security, and an embedded QWERTY physical keyboard. Blackberry was the device of choice by the government, many universities, and most businesses that required high security and inexpensive messaging.

However, in the late 2000s, Apple and Samsung began to out-innovate RIM. Apple created new ways for customers to use smartphones, such as an intuitive user interface and touchscreen navigation. Apple also developed means by which smartphones could make people's lives more convenient and more fun, and Samsung quickly followed suit. Apple saw the smartphone as more than just a communications device; they saw it as a component of a user experience system. Meanwhile, RIM did little to bring new features to its customers and failed to recognize the dynamic changes occurring in the business market.

Despite widespread consumer demand for hardware improvements (such as a touchscreen keypad, higher resolution, bigger screen, and faster Central Processing Units) and more applications (games, music, social media, interactive video, camera, and other entertainment), RIM focused instead on secure corporate communications. Well behind the competition, RIM finally launched a touchscreen device that was viewed as an inadequate imitation of the iPhone. By September 2013, the company announced a loss of almost $1 billion due to unsold inventory. Today RIM is a shadow of its former self, with capitalization of less than $4 billion (versus $82 billion in 2008) and a stock price down more than 90% from its high of $137.41 in mid-2008.

*Systems Thinking Lesson:* This is an unfortunate example of viewing products and services as stand-alone items and failing to recognize that most products are components of user experience systems. Blackberry continued to view its products as mobile e-mail devices, while Apple and Samsung created new ways to use smartphones as mobile entertainment devices, while continuing to provide e-mail services with an intuitive user interface.

4.1.4. Swiss Watchmakers

In the 1960s, the Swiss dominated the watchmaking industry, with over 60,000 employees and 65% of world-wide watch sales [45]. In those days, watches were analog devices with hands that were powered by windup mainsprings. In 1968, several Swiss researchers invented a new kind of quartz movement watch with no mainspring, powered by a tiny battery.

The Swiss watchmaking industry ignored this development in the belief that traditional watches would prevail forever—they did not even bother to patent the idea. In late 1968, the new quartz movement watch was displayed at the World Watch Congress. Seiko Japan saw the new product and immediately recognized its potential. They quickly began production of the new watches and started taking market share. By 1978 there were only 10,000 Swiss watchmakers and the Swiss had less than 10% of the watchmaking market. The industry was shaken again in 2015 with the introduction of Apple's smart watch, essentially a wrist-mounted computer that also happens to tell time. Stagnant mental models can ruin not only businesses but entire industries.

*Systems Thinking Lesson:* Failure to embrace a new mental model, or paradigm shift. Disruptive paradigm shifts, or changes in mental models, are common in business. Automobiles replacing horse-drawn carriages, the Internet displacing the library, CDs displacing vinyl records (and being displaced by online music), VHS tape being displaced first by DVDs and then by streaming movies, and personal smart phones displacing land lines are just a few examples.

## 4.2. Examples in Sales and Sales Management

### 4.2.1. Missed Sales Targets for the Introduction of New Products

Successful companies continuously introduce new products. However, sales targets for those products are often missed. Here is a typical scenario: Kate is a fantastic salesperson for the fictional Gooseberry Corp, which makes and sells smart phones, tablets, and laptops. She is top-notch and has exceeded her quota for each of the past four years. The company has just developed a new product, the "3D Tablet", which is more complicated, more expensive, and harder to sell than Gooseberry's other products. Kate's boss Rick wants Kate to push the new 3D tablet aggressively. Kate does not want to, arguing that this will reduce her sales of other products, decrease net revenue, and decrease her personal income. The situation has degenerated and Kate now thinks that Rick is obstinate and unreasonable. Rick thinks that Kate is insubordinate.

Kate is right to object; selling the unproven new product in lieu of the old will likely reduce her sales volume and hence her compensation. Removing Kate from her excellent performance selling established products to sell a new, unproven, complicated product is unwise. Instead, the company might consider hiring new, hungry salespeople or moving existing staff within sales or engineering to push the new product. This would provide sales growth without cannibalizing existing sales or compromising Kate's pay.

*Systems Thinking Lesson*: The sub-optimal structure of both the system for introducing new products and the compensation for selling them.

### 4.2.2. Sales Management

When a salesperson misses targets repeatedly, a good sales manager will analyze the salesperson's behavior and activities. In one such case, a sales manager observed declining sales for one of his salespeople. The manager also observed that the salesperson spent 70% of her time in the office, doing paperwork, scheduling appointments, etc. The manager sat with the salesperson and instructed her to get out of the office and visit customers and prospects at least 70% of the time. Over the following year, the salesperson was on the road 75% of the time, yet her sales did not increase. This is a classic Systems Thinking archetype called "seeking the wrong goal". Spending a lot of time with existing customers and unqualified prospects is not necessarily a good way to increase sales. Both prospects and existing customers should be qualified to determine if products would be beneficial, before asking salespeople to call on them.

*Systems Thinking Lesson:* If the company had already been presenting the salesperson with qualified leads, then this indicates either a defective Goals-Behaviors-Metrics-Rewards (GBMR) system (an incorrect desired behavior was established) or an unqualified or untrained salesperson. If the company had not been developing qualified leads, then an incorrect business structure existed.

4.2.3. The Sales Commission Structure

In many companies, salespeople are rewarded based on a regressive commission structure. The more they sell, the smaller percentage they receive. For example, a sale of $10,000 yields a commission paid of $1,000 while a sale of $100,000 yields a commission of $2,000. The result is a lot of low dollar-volume sales, low efficiency, and reduced profit. This very common structure is based on a mental model geared towards limiting salespeople's compensation, which is counter-productive. A progressive structure that pays salespeople a higher percentage of large-dollar sales might indeed enrich the salespeople, but would also enrich the company.

Similarly, in many companies, sales are paid commissions based on sales dollars, instead of on profit dollars. In response, sales people often reduce product price to increase sales volume. In response to this, the company imposes limits on the discounts that sales are permitted to offer, or hides the factory costs from the salespeople, or requires multiple approvals for each sale. This structure hamstrings the organization. A better approach would be to provide the salespeople full information on factory costs but reward them based on profit, not sales.

*Systems Thinking Lesson:* Poor structure deriving from poor mental models.

### 4.3. Examples in Pricing

#### 4.3.1. Bank of America's 2011 Decision to Charge $5 for Each Bank Card Transaction

In 2011, Bank of America instituted a charge of $5 each time a debit card was used to make a purchase. They figured that this would increase profits dramatically while negligibly impacting customers. They were wrong: customers objected violently and staged mass protests, such as "Occupy Wall Street" and "Occupy Boston" [46]. Bank of America quickly changed its position and eliminated the fee. This was a clear example of a company badly misjudging consumers' reactions to a new policy.

*Systems Thinking Lesson*: Unintended consequences of a business decision.

#### 4.3.2. Airlines' Decision to Charge for Checked Baggage

Starting in 2008, major U.S. airlines began charging customers for checked baggage. In reaction, passengers packed more and more into carry-on bags, filling overhead racks to bursting, thus creating a double inconvenience for travelers. This unintended consequence certainly contributes to consumers' hatred of the airline industry (according to Sheiresa Ngo of CheatSheet.com [47], airlines are the eighth most hated industry in the U.S.)

*Systems Thinking Lesson*: Unintended consequences of a business decision.

#### 4.3.3. The National Parks Service Decision to Increase Entry Fees

In 1998, the National Park Service was suffering a shortfall in covering operating expenses. They decided that a simple solution would be to raise entry fees, evidently not realizing that park attendance volume is a function of the entry fee (a balancing feedback loop). As a result, attendance dropped and the shortfall increased. (Note that this error was almost repeated by the Trump administration in 2018, but rejected at the last minute due to public outrage.) Assuming that purchase volume is insensitive to price is a classic linear thinking business error.

*Systems Thinking Lesson*: Unintended consequences of a business decision.

4.3.4. The Market Manager's Decision to Increase Product Prices by 25%

In a New England separations company in the 1990s, a market manager was responsible for setting prices for filtration products. At one review meeting, the company owner chided the manager for not having raised prices over the last five years. The market manager's reaction was to make up for those five years in one fell swoop by raising prices by 25%. He believed that customers would be grateful upon realizing that they had been enjoying a fixed price over the past five years, while the cost of living had increased 25%. This proved false. Customers deserted the company and sales volume fell precipitously. A systemic feedback loop shows that customers stop being customers when they feel they are being abused.

*Systems Thinking Lesson:* Unintended consequences of a business decision.

*4.4. Examples in Operations and Quality Assurance*

4.4.1. Poor Inventory Management (The Beer Game)

In many companies, inventory levels oscillate instead of remaining stable, negatively affecting carrying costs, stock-outs, and profitability. Similar large oscillations often appear in staffing levels, factory loading, and Research & Development project backlog. Common management practice is to blame the individuals involved for allowing such swings. However, it is typically inherent systemic delays that cause the oscillations; the individuals involved have little control or influence. The Beer Game [17] is a famous system dynamics model depicting how systemic structure and feedback loops with delays yield oscillation in systems. In the game, delays in order processing, shipping, and receiving cause large oscillations in beer inventories. Students playing the game initially blame the individuals making the purchase decisions, but eventually realize that the oscillations are the fault of delays in the system itself.

*Systems Thinking Lesson*: Failure to understand that systemic structure often impacts business results more than individuals do.

### 4.4.2. Quality Assurance Targets

Many companies establish annual performance goals at the corporate, departmental, and individual levels. In one company, the quality assurance (QA) department was goaled with reducing the defect rate in a key product. The QA Director met with the Production Supervisor to discuss the issue; the Production Supervisor in turn spoke with his shop supervisors. A shop floor supervisor then told his machine operator that quality is extremely important, and that his salary increase will depend heavily on quality improvement. Yet, for the following three months, the number of defects on the individual's machine remained the same. The machine operator really wanted to reduce the number of defects but did not know how.

*Systems Thinking Lesson*: A defective Goals-Behaviors-Metrics-Rewards (GBMR) system. Establishing employee goals without elucidating exactly how (behaviorally) to achieve those goals is unwise.

### 4.5. Examples in Administration

### 4.5.1. Major League Baseball in the 1990s

In the 1990s, the conventional wisdom in major league baseball was to pay huge bucks for big-name stars. Team payrolls varied by a factor of 3, and those teams with smaller payrolls had trouble fielding championship teams. In the 2011 Columbia Pictures movie *Moneyball*, this problem is exemplified when Oakland Athletics general manager Billy Beane is confronted with the challenge of assembling a winning baseball team on a limited budget. Beane's assistant, Peter Brand, points out that the huge dollar, superstar mental model is flawed and that there is greater value in using *sabermetrics* to identify and hire lower-priced players who may have been overlooked; advice that Beane embraces. Using the technique to focus on On-Base Percentage (OBP), Beane assembles a team that makes it all the way to the 2002 American League West title on a limited payroll. Other teams adopt the sabermetrics philosophy and two years later, the Boston Red Sox win the World Series using Beane's and Brand's novel mental model. This represents a paradigm shift in how baseball players are assessed and compensated.

*Systems Thinking Lesson:* An antiquated mental model with respect to how to assemble a winning team.

4.5.2. The New England Finned Fishing Industry's Decision to Take as Many Fish as Possible

Up until the 1980s, the finned fishing industry in New England was prosperous. When some noted that the fish stocks on George's Bank were being depleted and called for quotas on the amount of fish taken, the industry reacted violently, arguing that their survival depended on maximizing fishing hauls. There is a systemic feedback loop at work here: the more fish taken, the fewer remain for subsequent years. Of course, the unintended consequence was the rapid depletion of fish off New England waters and the concomitant demise of the industry; George's Bank was closed to fishing in 1994 and has not reopened.

This overlooked feedback loop is an example of the Tragedy of the Commons, in which competition for an underpriced resource results in its depletion as an unintended but inevitable consequence. Clearly, the fishermen did not understand the systemic implications of their practices. It is noteworthy that the Maine lobster industry, by using Systems Thinking and limiting fishing hauls, has remained stable and viable over the past 100 years.

*Systems Thinking Lesson:* Unintended consequences of a business decision.

4.5.3. The Common Business Decision to Cut R&D and Advertisements/Promotions When Sales Slip

A common business practice is to reduce Research and Development and Promotion/Advertising when sales slump. Of course, these functions are critical to the development of new products or services and the origination of new business; cutting them often leads to further reductions in sales. The result of this inattention to feedback is a short-term increase in profits and the unintended consequence of long-term losses.

*Systems Thinking Lesson*: Unintended consequences of a business decision.

### 4.5.4. Poor Justification for Hiring Additional Staff

Employees often feel that adding resources will increase profits but have trouble convincing their supervisors. In one case, a company had an excellent Regional Sales Manager (RSM) who was responsible for the western third of the U.S. He had two sales engineers and an administrative assistant working for him, and his region generated ~$6 million in sales each year. At every opportunity, however, he complained to his boss (the V.P. of Sales) that he was missing half the sales opportunities in his vast region because he did not have enough staff. The V.P. was not convinced that the extra staff would pay off, so he kept denying the RSM's request. The RSM and the V.P. grew frustrated with each other and sales volume remained flat. In this case, the problem was that the corporate structure had separated accountability from authority in a command-control structure.

One may view a business as either a system in which employees simply execute decisions made by management (a command-control structure, much like the military) or as a system in which employees are motivated properly to make good business decisions themselves (a market-based structure [48]). If the employee is given both authority to make resource decisions and accountability for making them (see Figure 4), and their compensation is based on their performance (a market-based management structure), then there is no need for a command-control structure, the role of the supervisor is changed from an approver to a coach/mentor/obstacle-remover, and interpersonal frictions are reduced.

*Systems Thinking Lesson*: A poor business structure based on a poor mental model.

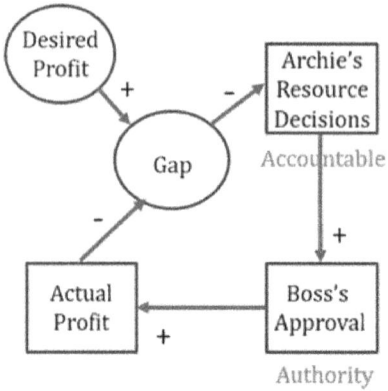

Command-Control Management      Market-Based Management

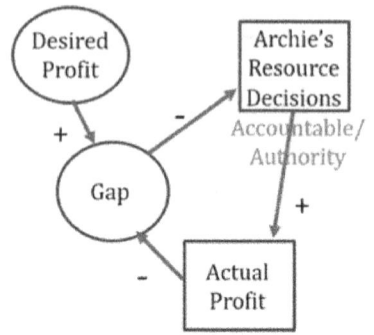

**Figure 4.** Systems Thinking in Hiring Accountability.

4.5.5. Conflicting Behavior Patterns

In some administrative situations, employees exhibit repeated patterns of conflicting behavior due to discrepant mental models. Consider the following scenarios:

i. Every time Rick walks into his boss's office to ask for help, she yells at him. Had this occurred only once, it might not have been a systemic issue, but its repetition indicates a pattern, which is caused by structure, which, in turn, is caused by an underlying mental model. In this case, the boss may not feel she is responsible to help Rick (a structural issue), or she may feel that Rick should know the answers for which he is seeking help (a mental model), or Rick may be intruding upon her at an inconvenient time (structure and mental model). The situation is not likely to improve until the two sit down together and discuss both of their mental models with respect to the support to be provided Rick and the correct procedures for soliciting it.

ii. Katie and Pete work on the same project team. In meetings and presentations, Katie repeatedly criticizes Pete's ideas, and Pete reacts negatively to the criticism. This pattern indicates an underlying structural issue that is engendered by mental models. Pete may not be following established procedures for implementing new ideas, or Katie may be overly assiduous in enforcing rules. Or, Katie may view Pete as a threat to her advancement within the company because corporate structure rewards people for new ideas. In any case, the situation may be addressed by analyzing the structure (rules, policies, procedures, etc.) and the underlying mental models held by both Katie and Pete as well as by their supervisor.

iii. Employees are required to travel occasionally on business, and reasonable business expenses are reimbursed via the company's expense report system. Recently, almost every expense report submitted by employees in Department X has been rejected by Accounting, which is a new behavior pattern. The employees are growing frustrated and are considering refusing to travel on business because of the difficulty getting reimbursed. There is suspicion that the accounting clerks have been instructed to reject expense reports specifically to reduce expenses and improve the company's profit. This situation will not be addressed until all parties understand the rules and procedures for expenses (structure) and the underlying mental models that have contributed to them.

*Systems Thinking Lesson*: Discrepant Mental Models yielding unclear or suboptimal business structures.

Table 2 offers a summary of business failures discussed with corresponding Systems Thinking Lessons.

**Table 2.** Overview of the Business Failures and Systems Thinking Lessons.

| Business Failure | Systems Thinking Lesson |
|---|---|
| *Product Development and Product Life Cycle Management* | |
| The Microsoft Zune | Failure to understand that many products are components of a *user experience system* |
| Polaroid | Flawed Mental Model |
| RIM's Blackberry | Failure to understand that many products are components of a *user experience system* |
| Swiss Watchmakers | Flawed Mental Model/Paradigm Shift |

## Sales and Sales Management

| | |
|---|---|
| Missed Sales Targets for New Products | Sub-Optimal Structure |
| Sales Management | A Defective Goals-Behaviors-Metrics-Rewards System/Sub-Optimal Structure |
| Sales Commission Structure | Flawed Mental Model/Sub-Optimal Structure |

## Pricing

| | |
|---|---|
| Bank of America's Decision to Charge for Transactions | Unintended Consequences |
| Airlines' Charge for Checked Baggage | Unintended Consequences |
| National Park Service Entry Fees | Unintended Consequences |
| 25% Price Increase | Unintended Consequences |

## Operations and Quality Assurance

| | |
|---|---|
| Poor Inventory Management | Sub-Optimal Structure |
| QA Targets | A Defective GBMR System |

## Administration

| | |
|---|---|
| Major League Baseball in the 1990s | Flawed Mental Model |
| Finned Fishing Industry Policies | Unintended Consequences |
| Cutting R&D and Ad/Promo | Unintended Consequences |
| Structure for Hiring Staff | Flawed Mental Model/Sub-Optimal Structure |

It is interesting and fortunate that the Systems Thinking failures explaining these 18 problems fall into just 4 categories:

1. Failure to understand that "products" are really components of user experience systems;

2. Flawed or outdated mental models/paradigms;

3. Structure and unintended consequences: failure to heed feedback in the system;

4. Flaws in the Goals-Behaviors-Metrics-Rewards system.

It is also interesting to note that all the failures in *pricing* fall into one category, unintended consequences: failure to heed feedback in the system.

In the subsequent section we further expand on these to offer a simple but useful protocol for applying Systems Thinking concepts in business.

## 5. The Systems Thinking Process/Protocol for Business

Having reviewed some of the Systems Thinking successes and failures with respect to business, it is possible to use inductive logic to develop some rules of thumb for the proper application of Systems Thinking principles to business and management. These are not intended to be sequential steps, but instead fundamental Systems Thinking principles that should be followed. The basic principles are:

1. Design and sell user experience systems, not products or services;

2. Expose, understand, and develop shared mental models/paradigms;

3.  Address structure and unintended consequences by identifying *feedback* in the workplace;

4.  Optimize the Goals-Behaviors-Metrics-Rewards system.

These four basic principles are detailed below.

*5.1. Design and Sell User Experience Systems, not Products or Services*

As Norman [42] says, a product is more than just a product; most products and services are merely components of *user experience systems* that involve the product's acquisition, delivery, packaging, use, support, maintenance, environment, disposal, and the product itself. Customers develop opinions based on all of these; a great product with poor maintenance and support is unlikely to be viewed favorably. Companies must take all these factors into consideration when producing products and services. So, when considering new products or services, companies should:

a.  Identify the user experience system in which the product/service fits;

b.  Identify all components of that user experience system. Include acquisition, delivery, packaging, use, support, maintenance, environment, and disposal, as well as the product itself;

c.  Ensure that all elements of the user experience system are focused on the user, not the company, and not the product;

d.  Structure the company to support this. Ensure that users will enjoy every part of the experience;

e.  Study Apple's development of the iPod user experience system for inspiration.


*5.2. Expose, Understand, and Develop Shared Mental Models/Paradigms*

Good mental models yield positive structures, and vice-versa. Mental models are the bottom level of the Iceberg Model (see Figure 1) and cause structures, behaviors, and patterns. For example, a mental model that incentive compensation increases productivity may yield a compensation structure that pays employees bonuses for new ideas or for exceeding targets. However, that same mental model and resulting structure may yield destructive competition and back-stabbing of colleagues. A mental model that prices must increase each year may result in tactical profit increases but long-term loss of business. Often, managers are not even aware of their own mental models and marvel at the company results caused by them. It is important to understand one's mental models and expose them to scrutiny, as well as others within (and outside of) the company. When business Key Performance Indicators do not meet expectations, managers should review their own mental models to see if they require modification. It is sometimes beneficial to have the management team sit together for the express purpose of reviewing corporate mental models. Mental models may include those focused on:

a. Who is the market and how big it is;

b. Attitudes of the market;

c. Perceived value of the good/service;

d. How to reach potential customers;

e. The selling approach;

f. Competition strengths and weaknesses;

g. Our own competitive advantages;

h. Employee motivation and incentive compensation;

i. Desired employee skills and experience;

j. The best way to produce and deliver the product or service.

*5.3. Address Structure and Unintended Consequences by Identifying Feedback in the Workplace*

Mental models engender business structures. Structure is the way that the system components interrelate in feedback loops and manifests itself as the company's rules, policies, procedures, authorities, and approval levels. Linear thinking fails to consider the many feedback loops inherent in businesses and, consequently, often results in unintended consequences. Although it is probably impossible to prevent ALL unintended consequences of business decisions, it should be possible to minimize them by understanding the systemic structures (specifically the feedback loops) that cause them. For example, raising prices increases short-term profit but negatively impacts sales volume and customer attitudes; consumption of resources increases short-term productivity but reduces the quantity of available resources in the future; reducing business origination systems saves money in the near-term but reduces future business. Linear thinkers often overlook the feedback loops, either inadvertently or *intentionally* to secure short-term profits at the expense of long-term consequences.

To avoid unintended consequences, identify the structures and the feedback loops in the company (see [49]). This can be accomplished using Systems Thinking tools, such as Stock-and-Flow diagrams and Causal Loop Diagrams (CLDs). For example, oscillations are almost always caused by feedback loops with *delays* [50]. In analyzing an oscillation in inventory levels, an analyst may develop the Causal Loop Diagram shown in Figure 5.

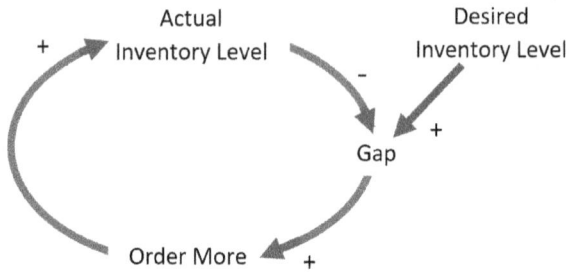

**Figure 5.** Inventory Causal Loop Diagram.

But this feedback loop would not yield oscillations. A more accurate CLD is shown in Figure 6, which depicts the delay.

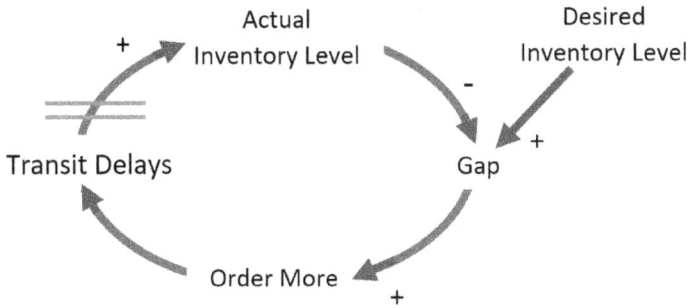

**Figure 6.** Corrected Inventory Causal Loop Diagram.

The solution is to first recognize that oscillations are caused by feedback loops with delays, then identify the delay(s), and then try to minimize or anticipate them.

It may be wise to schedule regular "unintended consequence" meetings to expose and discuss unintended results and trace back the underlying business structures and mental models that engendered them. Often, rules and procedures are amended and modified over time, leading to "Rube Goldberg" structures. In other cases, situations change and the policies are no longer current or valid. It is wise to review company rules, policies, procedures, protocols, incentives, and structures regularly, and to map out corresponding Causal Loop Diagrams, to ensure that they are yielding desired behaviors. Employees and customers affected by the policies and procedures should be included in these discussions.

Some of the more common causes of unintended consequences are pricing decisions, employee incentive structures, company reorganizations and structural changes, systemic delays, and resource utilization decisions. Monat and Gannon [49] provide several more examples of the application of Systems Thinking tools to correct unintended consequences in a sales organization.

*5.4. Optimize the Goals-Behaviors-Metrics-Rewards System*

One of the strongest systems thinking structures in any company is the Goals-Behaviors-Metrics-Rewards (GBMR) system, which is fundamentally a *feedback system* that positively reinforces employees for desirable behavior and negatively reinforces them for undesirable behavior, as shown in Figure 7.

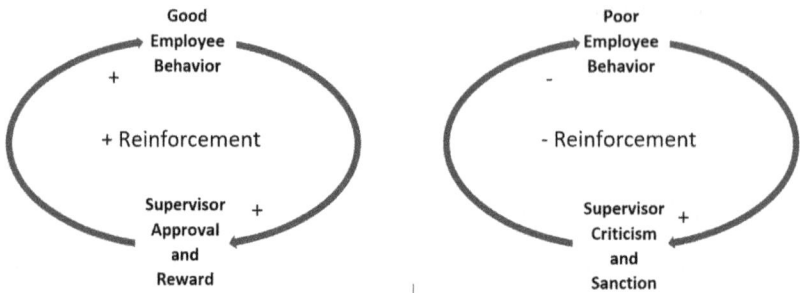

**Figure 7.** Employee-Supervisor Feedback Loops.

This system attempts to establish goals for employees and reward them for achieving those goals; ostensibly if all employees achieve their goals, the company will prosper. But many companies execute this poorly. In Reference [51], Monat asserts that many mangers articulate incorrect goals, fail to articulate desired behaviors, and establish metrics and rewards that are counter-productive.

**Organizational Goals** are hierarchical and start at the top of the company (Board of Directors, CEO, President) and percolate down through the organization. They should be different for each group and individual, with the lower level goals supporting the next level above. Goals should mesh and support each other both vertically and horizontally. They must be accurate: do not say "increase sales" if you mean "increase sales and profit"; do not say "improve delivery" if you mean "improve delivery without compromising quality and cost" (these are examples of the "seeking the wrong goal" archetype). Most (but not all) goals should be quantitative and most should be process-based as opposed to function-based (process-based goals penetrate across many divisions of a company, whereas function-based goals are within a division).

**Behaviors** are also hierarchical, like goals. Behaviors derive from goals, yet many managers fail to translate goals into desired behaviors, which describe what an employee must do to achieve the goals. Behaviors are developed jointly by the employee and the employee's supervisor. Note that two individuals with the same title may have different desired behaviors—they are tailored to the individual. Suppose a salesperson's goal is to increase sales volume by 10% over last year without lowering prices or providing more services for free. An agreed-upon behavior might be, "This year I will spend 50% less time dealing with customer issues (by passing them off to Customer Service) and 50% more time following up on screened leads provided to me by the Business Origination group. I will also contact every current customer listed in our Customer Relationship Management system to try to expand sales to them."

**Metrics**: Most goals are measured. However, it is important to also measure behaviors. Sometimes, these are hard to quantify. But ask, "How do I know if they are doing a good job?" If the desired behaviors have been articulated well, measuring them is straightforward. In the above example, it is a simple matter to measure, during the course of the year, if the salesperson is dealing with fewer customer problems and spending more time pursuing screened leads and contacting existing customers.

**Rewards:** Rewards are part of the systemic structure. Rewards are tools to encourage desired behavior and are components of the reinforcing feedback loop depicted in Figure 7. They should be within the company's financial constraints, immediately and clearly linked to the behavior that yielded the reward, significant to the employee (per Maslow's hierarchy), and at the discretion of the immediate supervisor (within reasonable bounds). Examples of good rewards include a case of gourmet foods (for a food connoisseur), a day off with pay, additional responsibility for someone who craves that, the opportunity to present a paper for someone who craves prestige and respect, a free dinner out for the employee and their spouse, and the ability to come in late after working extremely hard one week. Examples of bad rewards include commissions paid at year-end (the association between the behavior and the reward is too far separated), a $50 gift certificate for a million-dollar sale (the scale of the reward is incommensurate with the achievement), and a gold Cross pen and pencil set to an individual who can barely pay their bills (inconsistent with Maslow's hierarchy).

In summary, an optimized GBMR System has:

1. Clear and reasonable goals, based on hierarchy;

2. Behaviors defined by supervisors and their employees that, if effected, will lead to goal achievement;

3. Metrics that accurately measure behaviors, not goals;

4. Rewards that motivate.

For a more extensive discussion on an optimized GBMR system the reader is referred to [51].

## 6. Conclusions

Businesses are systems, the components of which include products and/or services, the physical building and its contents, employees, customers, stakeholders, management, the environment, regulatory agencies, banks, suppliers, communications vehicles, transportation services, performance metrics, and other factions. These components interact in complex and sometimes surprising ways. Systems Thinking is a perspective, a set of tools, and a language that may be used to understand and optimize system behavior. Yet, Systems Thinking is not used extensively in business management.

In this paper we discussed several business failures and identified Systems Thinking lessons that could be applied to avert such failures with the goal of communicating these concepts at the level of the practitioner. We synthesized these lessons into a simple, practical framework for Systems Thinking in business. The approach taken here was inductive and was not intended to follow the rigor of a deductive scientific theory. Future work building on this is proposed to validate and refine the proposed approach in the context of real case studies.

Many of the Systems Thinking concepts described in this paper are not new, but many practitioners fail to apply them in a business environment. We believe that this is because experience alone does not ensure that we do not repeat our mistakes. Disparate mental models, unintended consequences of decisions, poor employee rewards systems, and bad pricing decisions have all been observed and documented, yet those errors are repeated. To be useful, experiential learning must be organized into some convenient, logical format that is easily accessible: a paradigm. The Merriam-Webster dictionary defines "paradigm" as "a philosophical and theoretical framework of a scientific school or discipline within which theories, laws, and generalizations and the experiments performed in support of them are formulated." It would be much easier for business managers to avoid these types of mistakes if there were a Systems Thinking paradigm that could be applied as Standard Operating Procedure for businesses. It is our hope that the Systems Thinking business protocol described here constitutes this paradigm and will minimize these errors going forward.

**Author Contributions:**

All authors contributed to the conceptualization of the content for this paper. In addition, Dr. Amissah conducted the research for and prepared the initial draft of the literature review in Section 3, as well as the final formatting of the paper, including the citations to the references. Dr. Monat and Dr. Gannon conducted the research for and prepared the initial draft of the examples contained in Section 4, as well as the basic principles for the proper application of Systems Thinking principles to business and management in Section 5. All authors contributed to the introduction to Systems Thinking in Section 2, the conclusions in Section 6 and the review of the final draft of this paper.

**Funding:** This research received no external funding.

**Conflicts of Interest:** The authors declare no conflict of interest.

# References

1. Monat, J.P.; Gannon, T.F. What is systems thinking? A review of selected literature plus recommendations. *American Journal of Systems Science* **2015**, *4*, 11-26.

2. Senge, P.M. *The fifth discipline fieldbook: Strategies and tools for building a learning organization*; Crown Business: 2014.

3. Kim, D.H. *Introduction to Systems Thinking*; Pegasus Communications: Waltham, MA, USA, 1999.

4. Monat, J.P. *SYS 540 Introduction to Systems Thinking (Lecture Notes)*; Worcester Polytechnic Institute: Worcester, MA, USA, 2019.

5. Checkland, P. *Systems thinking, systems practice*; J. Wiley: Hoboken, NJ, USA, 1981.

6. Jackson, M.C. The origins and nature of critical systems thinking. *Systems practice* **1991**, *4*, 131-149.

7. Checkland, P.B.; Haynes, M.G. Varieties of systems thinking: the case of soft systems methodology. *System dynamics review* **1994**, *10*, 189-197.

8. Beer, S. *Cybernetics and management*; English Universities Press: London, 1959.

9. Beer, S. *Brain of the firm: a development in management cybernetics*; Herder and Herder: New York, USA, 1972.

10. Beer, S. *The heart of enterprise*; John Wiley & Sons: Hoboken, NJ, USA, 1979.

11. Beer, S. The viable system model: Its provenance, development, methodology and pathology. *Journal of the operational research society* **1984**, *35*, 7-25.

12. Schwaninger, M. Design for viable organizations: The diagnostic power of the viable system model. *Kybernetes: The International Journal of Systems & Cybernetics* **2006**, *35*, 955-966.

13. Jose, P.R. Design and diagnosis for sustainable organizations: The viable system method. Springer Science & Business Media: Berlin, Heidelberg, 2012; https://doi.org/10.1007/978-3-642-22318-1.

14. Forrester, J.W. Industrial Dynamics. A major breakthrough for decision makers. *Harvard business review* **1958**, *36*, 37-66.

15. Forrester, J.W. Urban dynamics. *IMR; Industrial Management Review (pre-1986)* **1970**, *11*, 67.

16. Forrester, J.W. *World dynamics*; Wright-Allen Press: Cambridge, MA, USA, 1971.

17. Sterman, J.D. Modeling managerial behavior: Misperceptions of feedback in a dynamic decision making experiment. *Management science* **1989**, *35*, 321-339.

18. Morecroft, J.D.; Asay, D.; Sterman, J.D. *Modeling for learning organizations*; Productivity, Incorporated: Shelton, CT, USA, 1994.

19. List of all cases. Availabe online: https://www.systemdynamics.org/list-of-all-cases (accessed on March 24, 2019).

20. Stacey, R.D. *Complexity and creativity in organizations*; Berrett-Koehler Publishers: San Francisco, CA, USA, 1996.

21. Stacey, R.D. *Tools and techniques of leadership and management: Meeting the challenge of complexity*; Routledge: New York, NY, USA. 2012.

22. North, M.J.; Macal, C.M. *Managing business complexity: discovering strategic solutions with agent-based modeling and simulation*; Oxford University Press: New York, NY, USA, 2007.

23. Checkland, P.B. Soft systems methodology. *Human systems management* **1989**, *8*, 273-289.

24. Scholes, J.; Checkland, P. Soft systems methodology in action. *Chichester, Wiley* **1990**, *876*, 910.

25. Maqsood, T.; Finegan, A.D.; Walker, D.H. Five case studies applying soft systems methodology to knowledge management. 2001. Available online: https://eprints.qut.edu.au/27456/1/27456.pdf (accessed on May 1).

26. Ackoff, R.L. Systems, messes and interactive planning. *The Societal Engagement of Social Science* **1997**, *3*, 417-438.

27. Beer, S. *Beyond Dispute: The Invention of Team Syntegrity*; Wiley: New York, NY, USA. 1994.

28. Flood, R.L.; Jackson, M.C. *Creative problem solving: Total systems intervention*; Wiley: New York, NY, USA. 1991.

29. Jackson, M.C. Creative problem solving: Total systems intervention. In *Systems methodology for the management sciences*, Springer: 1991; pp. 271-276.

30. Von Bertalanffy, L. General system theory. *General systems* **1956**, *1*, 11-17.

31. Kast, F.E.; Rosenzweig, J.E. General systems theory: Applications for organization and management. *Academy of management journal* **1972**, *15*, 447-465.

32. Forrester, J.W. Industrial dynamics. *Journal of the Operational Research Society* **1997**, *48*, 1037-1041.

33. Rebovich Jr, G.; White, B.E. *Enterprise systems engineering: advances in the theory and practice*; CRC Press: Boca Raton, FL, USA. 2016.

34. Churchman, C.W. *The systems approach*; Delacorte Press: New York, NY, USA. 1968.

35. Churchman, C.W. *The design of inquiring systems: basic concepts of systems and organization*; Basic Books: New York, NY, USA. 1971.

36. Mason, R.O.; Mitroff, I.I. *Challenging strategic planning assumptions: Theory, cases, and techniques*; John Wiley & Sons Inc: Hoboken, NJ, USA. 1981.

37. Eden, C. Cognitive mapping. *European Journal of Operational Research* **1988**, *36*, 1-13.

38. Bennett, P.G.; Huxham, C.S. Hypergames and what they do: A 'soft OR'approach. *Journal of the Operational Research Society* **1982**, *33*, 41-50.

39. Ulrich, W. *Critical heuristics of social planning : a new approach to practical philosophy*; Wiley: Chichester, 1994.

40. Senge, P.M. *The fifth discipline: the art and practice of the learning organization*, 1st ed. ed.; Doubleday/Currency: New York, NY, USA, 1990.

41. Ballé, M. *Managing with Systems Thinking: Making Dynamics Work for You in Business Decision Making*; McGraw-Hill International (UK): 1994.

42. Norman, D.A. The way I see it, Systems thinking: A product is more than the product. *interactions* **2009**, *16*, 52–54, doi:10.1145/1572626.1572637.

43. Smith, A.N. What was Polaroid Thinking? In *Yale Insights*, 2009. Available online: https://insights.som.yale.edu/insights/what-was-polaroid-thinking (accessed on May 1).

44. Gustin, S. The fatal mistake that doomed blackberry. *Time Magazine* **2013**. Available online: https://business.time.com/2013/09/24/the-fatal-mistake-that-doomed-blackberry/ (accessed on May 1).

45. Barker, J.A.; Stiever, G. *Joel Barker's The New Business of Paradigms*; Star Thrower Distribution: 2001.

46. Moore, G. Occupy Wall Street: One. Bank of America: Zero. Boston Business Journal 2011. Available online: https://www.bizjournals.com/boston/blog/bottom_line/2011/11/occupy-wall-street-one-bofa-zero.html (accessed on May 1).

47. Ngo, S. This Is the No. 1 Most Hated Industry in America, According to Real Customers; 2018. Available online: https://www.cheatsheet.com/money-career/most-hated-industries-in-america-according-to-customers.html/ (accessed on March 25).

48. Gable, W.; Ellig, J. *Introduction to market-based management*; Center for the Study of Market Processes: Fairfax, VA, USA, 1993.

49. Monat, J.P.; Gannon, T.F. *Using Systems Thinking to Solve Real-World Problems*; College Publications: London, UK, 2017.

50. Monat, J. Explaining natural patterns using systems thinking. *American Journal of Systems Science* **2018**, *6*, 1-15.

51. Monat, J. The integrated approach to optimizing productivity. *WorldatWork Journal* **2005**, *14*, 61.

# FAILURES OF SYSTEMS THINKING IN U. S. FOREIGN POLICY

Jamie P. Monat and Thomas F. Gannon
Systems Engineering Program
Worcester Polytechnic Institute
Worcester, MA, USA 01609

*reprinted from American Journal of Systems Science*, Vol. 5 No. 1, 2017, 1-12

I.      Abstract

Systems Thinking can be used to analyze and solve complex real-world problems that cannot be solved using short-sighted linear thinking. It can also help to understand complex international issues, such as the rise of terrorism and support for anti-American activities around the world; and to understand the illogical behaviors of organizations such as ISIS. In this paper, we apply the Systems Thinking methodology described by Monat and Gannon (2017) to analyze America's foreign policy approach over the past 40 years. We conclude that the United States' foreign policy has failed to use Systems Thinking in dealing with international issues. Instead of a cohesive strategy, the foreign policy has been one of short-sighted tactics, often with dire consequences. Examples include the invasions of Iraq and Afghanistan, the 2011 invasion of Libya, the arming of the Mujahedeen in Afghanistan, and even the rise of ISIS. Fundamental System Thinking principles that have been absent in addressing international issues include failure to recognize unintended consequences, failure to recognize and understand feedback loops, fixes that fail, poor root-cause analysis, and seeking the wrong goal. These failures are not exclusive to any one administration, but seem to be part of a pattern whose roots are embedded in the cultures of the U. S. State Department, the military, and the intelligence community.

II.    Introduction

After WW II, U. S. foreign policy seemed to embody a coherent strategy for dealing with the rest of the world. Characterized by liberation of invaded countries, economic development, and an interest in world peace, the U. S. philosophy was not tainted by ulterior motives such as the need for oil or the desire to act as the world's policeman. This changed gradually over the ensuing years to the point where U. S. actions have often done more harm than good. The 2003-2011 Iraq war, for example, has been cited as "a reason for the diffusion of jihad ideology" (Mazzetti, 2006; National Intelligence Council, 2006). This damage is typically the result of short-sighted linear thinking, in which feedback loops are either not identified or ignored, the systemic root cause of issues is not discovered, and unintended consequences predominate. Much of this damage could be avoided (and international relations set on an appropriate course) through the application of Systems Thinking to U. S. foreign policy.


III.    What is Systems Thinking?

Systems Thinking is a perspective, a language, and a set of tools that can be used to address complex political and socio-economic issues. Systems Thinking is the opposite of linear thinking. It is a holistic approach to analysis that focuses on the way a system's constituent parts interrelate, and how systems work over time and within the context of larger systems. Systems Thinking allows one to recognize that repeated events or patterns are derived from systemic structures which are derived from mental models. It also helps one recognize that behaviors are derived from *structure*. Systems Thinking focuses on relationships rather than components, considers the short and long term consequences of actions, recognizes the emergence of unintended consequences, and recognizes the principles of self-organization. Systems Thinking tools include system archetypes, causal loops with feedback and delays, and systemic root cause analysis. For a comprehensive summary of systems thinking terms, tools and techniques, see Monat and Gannon (2015).

System archetypes are patterns of behavior that are repeated in a variety of situations and organizations. They include "accidental advisories", "fixes that fail" and "seeking the wrong goal." The "accidental adversaries" archetype can occur when two or more entities initially either collaborate or exist harmoniously. Eventually one entity does something that the other entity perceives to be damaging, so that entity reacts. In turn, the other entity reacts, and the pattern of behavior results in a death spiral. The U. S. and North Korea are an example. The U. S. has no real interest in North Korea (either positive or negative) but views its missile development and nuclear program as a threat. It reacts by conducting military exercises with South Korea in the Yellow Sea and Sea of Japan. North Korea views these (along with the U. S.'s anti-communist philosophy and friendship with South Korea) as threats, and so ramps up its military programs; hence the death spiral. In many cases, the problem is that the actions of a partner are viewed as adversarial, when the partner is only guilty of pursuing a self-interest without taking into account the effects of those actions on the other partner. Local thinking, mistrust, and poor communication contribute to this archetype.

The "fixes that fail" archetype is another common behavioral pattern which identifies a problem and attempts to fix it, only to find out that the "fix" causes another problem. This archetype typically occurs when trying to "fix" problems in political, organizational or social systems that have very strong stabilizing or balancing feedback loops. Examples include the prohibition of alcohol in the 1920s and the overthrow of Sadam Hussein, which unleashed ethnic resentment between Iraqi Sunnis and Shiites based on competition for power, and ultimately gave rise to ISIS as an unintended consequence. Another example is the decision by the State Department to arm Sunni rebels against Assad in Syria, only to provide ISIS with the opportunity to seize billions of dollars in U. S. military equipment after capturing Mozul as another unintended consequence.

The "seeking the wrong goal" archetype arises when one establishes a goal which may be easier to accomplish or measure, but does not represent the desired end result. One example is the goal of U. S. foreign policy to act as the world's policeman to achieve world peace, which has resulted in the rise of terrorism, resentment from other countries, and support for anti-American activities around the world. Another example is the military's objective of incapacitating terrorists. This addresses a symptom, not the problem itself, which is poverty, oppression, corruption, and despair. We may eventually disarm or kill all ISIS, Al Qaeda, and Taliban terrorists currently active in the Middle East. But until we address the root causes that give rise to terrorism, new terrorist groups will rise under different names.

Causal Loop Diagrams (CLDs) show how various system components inter-relate and interact with each other. They are useful in illustrating feedback processes, which are present in most systems and often ignored in linear thinking analyses. Examples include stabilizing feedback loops, such as a home thermostat, and reinforcing feedback loops, such as interest compounding in a bank account. Another example of a reinforcing feedback loop is the continued intervention by the U. S. in local conflicts, which reinforces the resentment of other countries toward the U. S., gives rise to increased anti-American activity, and results in more U. S. intervention as depicted in the Causal Loop Diagram of Figure 1.

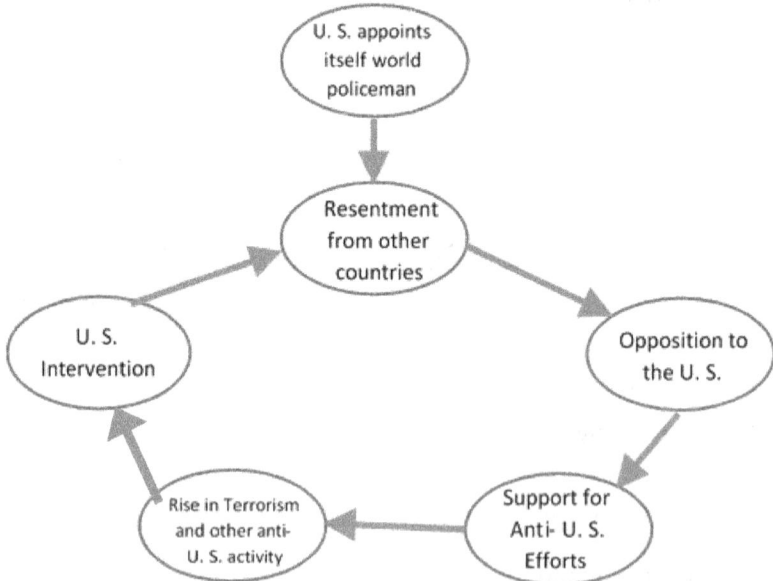

Figure 1. Causal Loop Diagram Showing the Impact of the U. S. Acting as the World's Policeman

Root cause analysis is a problem solving method focused on determining the fundamental, or "root" cause of a problem. Tools such as "The Five Whys" and fishbone cause-and-effect diagrams are used to facilitate the analysis of the root causes of a problem. Conventional root-cause analysis often ends prematurely, with root causes such as "operator error," "mechanical failure," "design fault," or "insufficient training" (Galley, 2014). Although these are valid, they are examples of linear thinking and do not address the underlying system or structure that gave rise to those causes.

IV.    Methodology and Approach

Monat and Gannon (2017) present a good methodology for applying Systems Thinking to solving real-world problems. They argue that "Solving a problem using Systems Thinking begins with stating the problem or issue, defining the system, applying appropriate tools, and drawing conclusions. Those tools must be selected and the optimal sequence of application must be customized for each specific situation." Their approach is summarized here:

Step 1. Develop and articulate a problem statement.

Step 2. Identify and delimit the system.

Step 3. Identify the Events and Patterns.

Step 4. Discover the Structures.

Step 5. Discover the Mental Models.

Step 6. Identify and Address Archetypes.

Step 7. Model (if appropriate).

Step 8. Determine the systemic root cause(s).

Step 9. Make recommendations.

Step 10. Assess Improvement.


They note that for a specific problem not all steps need be followed (Step 7 -- Dynamic Modeling, for example, may not be useful in certain situations) and that the sequence of steps may be changed as appropriate.

Step 1: Problem Statement: The U. S. foreign policy approach and actions over the past 40 years have often done more harm than good, resulting in an increase in terrorism, loss of lives, wasted resources, economic and humanitarian hardship, and loss of respect for America in the world.

Step 2: System Definition and Boundaries: The system to be analyzed is the U. S. State Department, military, and intelligence community's foreign policy culture, philosophy, approach, and actions with respect to terrorism originating in the Middle East and Northern Africa. The Middle Eastern/Northern Africa people, governments, cultures, and attitudes are a part of this system.

In the following examples, we apply steps 3-6 and 8-9 of the methodology described above and demonstrate how the lack of or poor application of fundamental Systems Thinking principles can result in short-sighted tactics with dire consequences in addressing international issues. Step 7 was deemed inappropriate for this analysis and step 10 remains for future consideration.

V.    Examples

**Arming of Jihadi Rebels in Afghanistan (1979-89)**

The Soviet Union invaded Afghanistan in 1979 to support the Communist regime of the People's Democratic Party of Afghanistan, which was losing popular support to anti-Soviet Islamic tribal factions outside of Kabul (U. S. Department of State, 2017). By the time of the invasion, the U. S. (under President Jimmy Carter) had already begun supplying the Mujahedeen, or Afghan Muslim freedom fighters, with non-lethal aid, to bolster their battle against communism and Soviet dominance. The Soviets were hoping for a quick, decisive victory; instead the war would drag on for 10 years and cost billions of dollars and millions of lives.

The U. S. saw the war as an opportunity to defeat (or at least restrict) Communism and the Soviet Union (vestiges of the debunked *Domino Theory*); radical Islamic terrorism was not a concern. The Mujahadeen were viewed as freedom fighters, and during the Reagan administration (1980-1988) the U. S. supported the Mujahadeen with billions of dollars, weapons, and training (which included the use of car bombs, assassinations, and terrorism (Gane-McCalla, 2011)). (The movie *Charlie Wilson's War* glorifies the efforts of Texas Congressman Charlie Wilson to secure weapons for the oppressed Afghan rebels.) The support of the rebels was easy to rationalize with a flawed mental model: they were a faithful religious faction fighting a godless Soviet invader and thus supported traditional American values. Zbigniew Brzezinski told the Mujahadeen, "We know of their deep belief in God, and we are confident their struggle will succeed. That land over there is yours, you'll go back to it one day because your fight will prevail, and you'll have your homes and your mosques back again. Because your cause is right and God is on your side." Eventually, they did prevail, and the Soviets were driven out of Afghanistan having failed to unite the country under Soviet rule.

Several unintended consequences developed as a result of U. S. actions. During the 10-year war, several Mujahadeen factions grew in both number and power: the Taliban, Al Qaeda, and the Muslim Brotherhood. Armed by the U. S., the forerunners of these factions were initially allied with America. But as the jihad strengthened in the late 1980s, so did anti-American sentiment. Several jihadist leaders such as Gulbuddin Hekmatyar (who had received $600 million from the U.S) and Abdul Sayyaf became more openly hostile toward the U. S. and initiated a propaganda war against not only the USSR, but also the U. S. The rising anti-U. S. sentiment was largely ignored by the second Regan administration as well as by the first Bush administration, neither of which chose to study or understand the local culture and systemic root causes of the anti-American sentiment (Coll, 2004).

One of the Mujahadeen was a young, wealthy Saudi named Osama Bin Laden. British Foreign Secretary Robin Cook said, "Bin Laden was, though, a product of a monumental miscalculation by western security agencies. Throughout the 80s he was armed by the CIA and funded by the Saudis to wage jihad against the Russian occupation of Afghanistan" (Gane-McCalla, 2011).

Eventually Bin Laden, Al Qaeda, and the Taliban became among the most notorious terrorist organizations on earth. And the U. S. had funded, armed, and trained them.


*Systems Thinking Gaffs: the Mujahedeen in Afghanistan*

- Failure to understand Systemic Root Causes:
    - The U. S. had learned about the difficulties of waging an ideological war thousands of miles from home from Vietnam; however it still did not appreciate the need to understand local politics, values, and psychological needs. As Steve Coll says, "The idea that Afghanistan was a messy place filled with complexity and ethnicity and tribal structures and all of the rest of what we now understand about Afghanistan was it was generally not part of American public discourse."

    - The Mujahedeen didn't hate the USSR because they were communists; they hated them for the same reason they would come to hate the U. S.: imperialism and a wide gap in values. Failure to understand basic Mujahadeen values.

- Erroneous Mental Models:
    - The enemy of my enemy is not necessarily my friend; and in retrospect a U. S. alliance with the USSR against the Mujahedeen might have served the world better.

    - Just because a group believes in God and doesn't like Soviet influence doesn't mean that the *bases* for those beliefs are consistent with ours.

- Unintended consequences:
    - Growth of the Taliban, Al Qaeda, and the Muslim Brotherhood

    - Growth of Osama Bin Laden's power and influence

## Al Shabaab in Somalia (2000-present)

After the overthrow of dictator Siad Barre in 1991, Somalia was a violent, chaotic place with severe food shortages. Local warlords and militias brutalized rival factions and killed and tortured ruthlessly (Adow, 2008, Hogg, 2008, Munger, 2015). These militias often aligned with local "courts" which maintained their versions of Islamic Sharia law. In 2000, several of the local courts united to form the Islamic Courts Union (ICU) in Mogadishu. The ICU provided health services, education, and security as well as law, and the areas they controlled became safer than warlord-controlled regions; their popularity grew steadily in the early 2000s (James, 1995; Stanford, 2016b).

But the U. S. viewed the ICU as an extremist Islamic group and feared that Somalia was becoming a haven for terrorists. Acting as the world's policeman, in 2006 the U.S. supported the development of the *Alliance for the Restoration of Peace and Counter-Terrorism* (ARPCT), an unlikely coalition of unpopular warlords who were not affiliated with the ICU (Bureau of Investigative Journalism, 2012). This generated substantial local resentment and instead of helping defeat the ICU, resulted in greater support for it; a major unintended consequence. Riding this wave of popular support, Al Shabaab, the ICUs military wing, defeated ARPCT in 2006 and as a result gained power and influence.

Meanwhile, Somalia's "official" government, the *Transitional Federal Government* (TFG) was attempting to consolidate power outside Mogadishu in opposition to the ICU. With U. S. and U. N. support, Ethiopia entered into the fray on the side of the TFG. In late 2006 the U. S.-supported Ethiopian army defeated the ICU and took control of Somalia. An unintended consequence of this action was the separation of Al Shabaab from the ICU as an independent military organization that became identified as the resistance against the occupying Ethiopian forces. Al Shabaab developed an alliance with Al Qaeda in a broad jihadist movement that embraced terrorist attacks on civilians (Associated Press/CBS, 2012.)

There have been various changes in Al Shabaab leadership and alliances since 2006, and the organization lost some local support when Ethiopian troops withdrew from Somalia in 2008. However, Al Shabaab continues as a radical Islamic terrorist group and has been responsible for several suicide bombings in Uganda and Kenya and for attacks on local relief workers in Somalia (Stanford, 2016a). Somalia is more stable now than in 2000-2012; however the war between the government and Al Shabaab continues in 2017 and Al Qaeda has been strengthened as a result of its association with Al Shabaab (another unintended consequence).

*Systems Thinking Gaffs: Al Shabaab in Somalia*

- Unintended consequences:
    - The rise of local support of the ICU *was a result of* U. S. support of ARPCT, and contributed to the resulting rise of Al Shabaab.
    - The U. S. support of the Ethiopian invasion strengthened both Al Shabaab and Al Qaeda.
- Feedback loops:
    - The invasion and defeat of the ICU caused Al Shabaab to separate.
    - The "U. S. as the world's policeman" mental model generated local anti-U. S. sentiment. The U. S. was viewed as an invader.

## Afghanistan (2001-present)

Afghanistan is a complex country, characterized geographically by rugged mountains, caves, and a harsh climate. This makes it especially beneficial for the locals who used the terrain to hide and organize during conflicts. And there have been many conflicts among the various tribal factions and invaders, over hundreds of years. The country is approximately 42% Pashtun (a tribal people who strongly identify with clans and follow "Pashtunwali," a self-governing tribal system) and 27% Tajik (a Persian-speaking group of Iranian descent who are not organized by tribes (Afghanistan Language and Culture Program (ALCP), 2017; Asian Wall Street Journal, 2011; Gouttierre and Baker, 2003.)) Administratively, there has never been a strong centralized government; instead rule is maintained by local Pashtun commanders in many areas, especially the northeast. Afghanistan is one of the poorest, most corrupt countries in the world (Chapman, 2016). The poppy-based drug trade has yielded profits ranging from 13-60% of the country's GDP, 23% of which is consumed by bribery (McCoy, 2016; UNODC, 2010). Drug trafficking provides substantial funding for the Taliban.

After the Vietnam War disaster and the similar Soviet debacle in Afghanistan, one would think that the U. S. would have learned its lesson: that ideological wars fought thousands of miles away without an understanding of local factions, cultures, values, and psychological needs lead to disaster; that the enemies of our enemies are not necessarily our friends; and that serving as the world's policeman is not viewed favorably. Such was not the case, and those erroneous mental models persist.

Following the September 11, 2001 attacks, the. U. S invaded Afghanistan with the objectives of driving out the Taliban, dismantling Al Qaeda, and capturing or killing Osama Bin Laden (BBC, 2017; Witte, 2017.) By late 2001, the Taliban had been driven from power, American ally Hamid Karzai had been installed as the Afghan leader (later president,) Bin Laden was on the run, and the U. S. led victory seemed clear.

But in a series of unintended consequences, most Taliban and Al Qaeda fighters had escaped to neighboring Pakistan or to the rugged, mountainous regions of Afghanistan, and Taliban leader Mohammed Omar regrouped and subsequently led an insurgency against the official Afghan government and U. S. allies. With the Taliban driven out, the Pashtun regional warlords regained local control, resulting in a chaotic mix of small tribal fiefdoms. And what seemed like a quick, decisive victory turned into a drawn-out, 16 year war (the longest in U. S. history) with a concomitant loss of 2,300 American lives at a cost of more than $800 billion (Chapman, 2016). Our use of superior military strength and faith in the righteousness of our cause failed to consider the inevitable feedback loops and again paved the way for disaster.

Today the Taliban are alive and well in Afghanistan (Almukhtar, 2017, Roggio, 2017, Qazi, 2017, Mitchell, 2017, Australian Broadcasting Corporation, 2017). Suicide bombings and terrorist attacks are common and the Taliban have regained control of ~30-40% of the country (see Figure 2). This is partly because of the flawed U. S. mental models that Afghans would appreciate both being "freed" from their Taliban oppressors and the installation of a democracy; that they would understand that a few civilian deaths are inevitable; and that destroying the drug-based source of Taliban funding would be wise. Acting on these assumptions, The U. S. destroyed poppy fields, which happened to be the source of many rural Afghan's income. Extensive civilian deaths were caused (as collateral damage) during allied air strikes. It is no wonder that we have lost the "hearts and minds" of most Afghans.

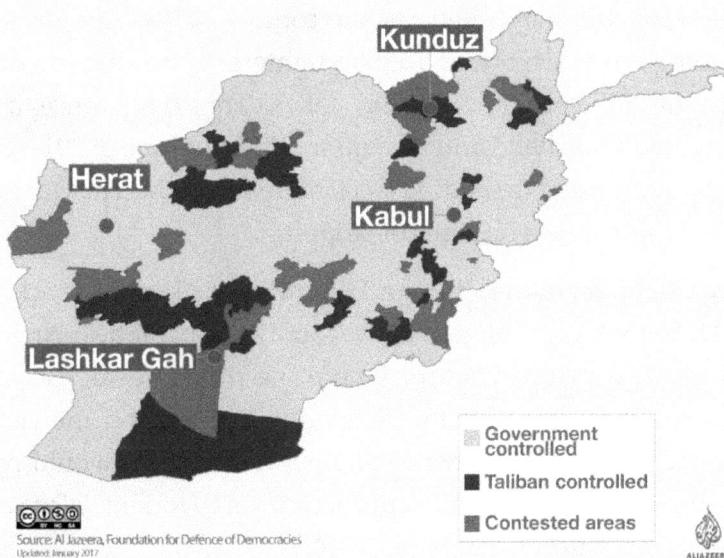

## Afghanistan: Who controls what

Kunduz

Herat

Kabul

Lashkar Gah

- Government controlled
- Taliban controlled
- Contested areas

Source: Al Jazeera, Foundation for Defence of Democracies
Updated: January 2017

ALJAZEERA

Figure 2. Taliban Influence in Afghanistan (from Qazi, 2017)

The U. S.'s biggest enemy in Afghanistan is not the Taliban—it's the Afghan way of life. In a culture dominated by illegal drug trade and corruption, American ideals and values do not hold. And assuming that a drug-free, democratic society would win the "hearts and minds" of the locals reveals a fatal misunderstanding of Maslow's hierarchy of human psychological needs (Maslow, 1943). As the U. S. foreign policy has revealed repeatedly, local populaces are not interested in political ideals and freedom of expression when they are starving and being killed by air strikes.

Millions of Afghans have fled the bloodshed, hunger, and corruption and emigrated as refugees to Pakistan, Europe, and Iran. Unemployment is at 40%. Drug profits range from 13-60% of the GDP. The official Kabul government regularly requests billions of dollars in charity. But most donated funds are squandered: *Transparency International* rates Afghanistan as one of the most corrupt nations on earth (Chapman, 2016). Why would the people support a government subject to bribes, simony, nepotism, and self-enrichment?

The U. S is not winning the war in Afghanistan, according to Defense Secretary James Mattis, who argues that we need a new strategy (Daniels, 2017). The Taliban are popular – they support the local way of life --- and may even regain control of the country. ISIS (in collaboration with former Taliban fighters) is making inroads in the north of the country and has taken Tora-Bora, once a Taliban stronghold (Nordland and Abedjune, 2017.) In June of 2017, President Trump committed to sending 4,000 more troops in. The war has lasted 16 years and shows no signs of stopping.

Steve Chapman of the Chicago Tribune says, "Looking back, it's clear the U. S. took on a project far beyond its capabilities. Arriving in a backward, war-torn country where we didn't speak the language, know the history, share the religion or understand the culture, we assumed that Marines, money and good intentions would produce a happy outcome. So far, that approach hasn't worked. What makes us think it ever will?" (Chapman, 2016).

*Systems Thinking Gaffs in Afghanistan*

- Failure to appreciate feedback loops: the reaction of the Taliban to the U. S. invasion

- Failure to do systemic root cause analysis and understand local factions, cultures, values, and psychological needs

- Poor Mental models:

  o  That conversion to a democratic society would win the "hearts and minds" of the locals reveals a fatal misunderstanding of Maslow's hierarchy of human psychological needs.

  o  Failure to understand that the drug trade (poppy farming) is the basis for many civilians' survival yielded resentment.

  o  "Democracy is the best form of government" and "The U. S. is the world's policeman" caused oppression of minorities and resentment by many.

- o Failure to understand local mental models (e.g. poppy cultivation is the key to feeding and sheltering my family) and the structures that they engender.

- o Erroneous mental model that superior military strength and fighting for a righteous cause inevitably lead to victory. An unintended consequence of this mental model is that our opponents have figured out how to fight a war knowing that they are vastly outgunned. In that sense, one could argue that the vast military strength of the United States has promoted terrorism.

## Invasion of Iraq and the Rise of ISIS (2003 - Present)

In March 2003, the U. S. led an invasion of Coalition forces into Iraq, which eventually ended the twenty-four year reign of Iraqi President Saddam Hussein. According to General Tommy Frank, the objectives of that invasion were to end the regime of Saddam Hussein; to identify, isolate and eliminate Iraq's weapons of mass destruction; search for, capture and drive out terrorists from that country; collect intelligence related to terrorist networks; collect intelligence related to the global network of illicit weapons of mass destruction; immediately deliver humanitarian support to the displaced and many needy Iraqi citizens; secure Iraq's oil fields and resources, which belong to the Iraqi people; and help the Iraqi people create conditions for a transition to a representative self-government (Sale, 2003). While the existence of weapons of mass destruction and a global network of illicit weapons of mass destruction were never found after extensive inspections and intelligence operations, there was clearly a need to provide humanitarian support to many needy Iraqi citizens and help the Iraqi people create conditions for a transition to a different type of government. Unfortunately, the U. S. decided that democracy was the best form of a representative self-government for Iraq. To that end, a permanent 275-member Council of Representatives was elected in the December 2005 by an Iraqi parliamentary election, which formed the Government of Iraq in 2006. Additional parliamentary elections were held in 2010 and 2014, which increased the membership of the Council of Representatives to 325 and 328, respectively, who in turn elected the Iraqi President and Prime Minister. The continued expansion of membership in the Council was intended to provide a growing list of political parties with a voice in government affairs. But "democracy" does not mean the same thing to everyone. To people who have been oppressed by a dictatorship, "democracy" may mean that the majority can now make rules that subjugate all minorities. The 2010 election created a large Shia bloc, which created an authoritarian regime to the disadvantage of the other political parties, most notably the Sunni political regime (BBN News, 2011). According to *Transparency International*, Iraq (along with Afghanistan) is one of the most corrupt governments in the Middle East (Agator, 2013), and is described as a "hybrid regime" between a "flawed democracy" and an "authoritarian regime." A 2011 "Costs of War" report from the Watson Institute for International Studies at Brown University concluded that U. S. military presence in Iraq has not been able to prevent this corruption, noting that as early as 2006, "there were clear signs that post-Saddam Iraq was not going to be the linchpin for a new democratic Middle East" (Balagi, 2011).

The ongoing resentment between competing regimes within the Iraqi government gave rise to a vicious cycle of oppression between the Sunni and Shia political regimes, which ultimately led to the failure of the Iraqi government and a June 2014 attack on Mosul by ISIS and insurgents led by Abu Abdulrahman al-Bilawi. During this period, ISIS popularity in the region continued to grow due to ISIS's provision of free housing, food, clothing, and health care for all, especially with respect to Sunni Muslims who had not enjoyed those benefits from their government. In addition, a growing number of unemployed youth in Iraq perceived their government as corrupt and authoritarian, and viewed ISIS to be more fair, just, and moral, as well as providing an opportunity for a better life.

Imposing democracy as a representative form of self-government in Iraq without understanding the region's culture, history, and politics is an example of a "fix that failed." This ultimately led to the rise of

ISIS as an unintended consequence of the U.S. intervention in Iraq, as illustrated in Figure 3.

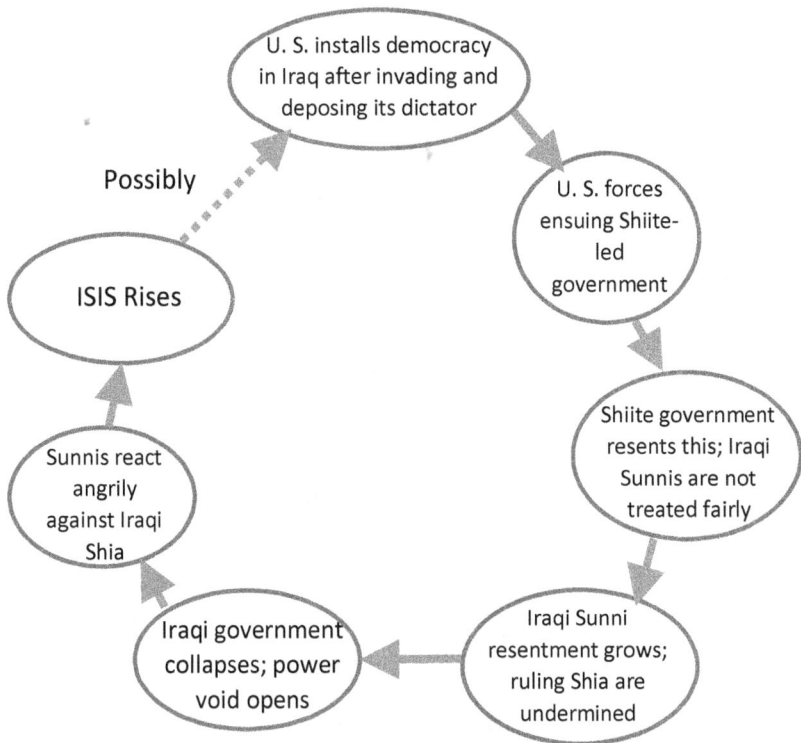

Figure 3. Causal Loop Diagram Depicting the Rise of ISIS

This mistake could be repeated in the future if subsequent U. S. intervention is taken to defeat ISIS and impose democracy again without learning from this previous gaff.

*Systems Thinking Gaffs in the Invasion of Iraq and Rise of ISIS*

- Failure to execute Systemic Root Cause Analysis to understand the bases for ISIS appeal: understanding of local culture, history and politics.

- Unintended consequence: the Rise of ISIS.

- Mental models that democracy is the best form of government and that the U. S. is the world's policeman.

- Failure to appreciate feedback loops: the oppression of competing political regimes and "fixes that fail."

## 2011 Invasion of Libya and Resultant Turmoil in Mali

In 2011, Libya was immersed in a civil war. Forces loyal to President Muammar Gadaffi clashed with opposition forces seeking to overthrow his brutal, totalitarian regime. Civilian casualties were high. To prevent a "humanitarian disaster," Secretary of State Hillary Clinton, U. N. Ambassador Susan Rice, and Samantha Power and Gayle Smith of the National Security Council convinced President Obama that military intervention was necessary (Kinzer, 2012). The result was the passage of U. N. resolution 1973 which called for an immediate ceasefire and end to all violence against Libyan civilians. In March of 2011 a NATO coalition led by American forces began missile launches, air strikes, a naval blockade, and enforcement of no-fly zones. Gadaffi was killed in October and governmental authority was assumed by the Transitional National Council of Libya (NTC).

As often occurs with military interventions, there were unintended consequences. Gadaffi's army consisted of a large number of Tuaregs: Islamic nomads from Mali. After the war, the Tuaregs returned to Mali with weapons, experience, and training, and attempted to establish a new homeland in northern Mali. This militant group overran the weak Mali army which had at least maintained a semblance of stability. With the Mali army in disarray and the Tuaregs clamoring for a homeland, the resulting chaos attracted Al Qaeda and other Islamic extremists (some of whom had been trained and armed by Americans (Iheduru, 2012)) who saw an opportunity to spread the rigid Salafi form of Islam, destroying cultural icons and tombs, looting schools and clinics, and exacting harsh punishment from those who do not follow the letter of Salafi law (Kinzer, 2012). Various competing jihadist groups, skirmishes, and failed peace agreements have plagued the region since 2012. Today, Mali is a land of corruption, drugs, extremists, suicide bombings, and narco-terrorism (Washington, 2016). The sale of illicit drugs finances terror campaigns, and the Associated Press states that "The U. N. mission in Mali is the deadliest active peacekeeping mission in the world."

Stephen Kinzer (2012) says, "Intervening violently in the politics of another country is like releasing a wheel at the top of a hill: you have no idea where it will end up. Perhaps it is too much to expect that well-meaning amateurs...would know enough about the country to understand what the consequences of their actions might be. It should at least be possible, however, to hope that policy planners would recognize their ignorance. A dose of humility might lead them to realize that military intervention always produces unforeseen consequences...... Overwhelming military power guarantees short-term victory....No amount of weaponry, however, can prevent the devastating "blowback" that often follows." The "blowback" to which Kinzer refers is the inevitable feedback or reaction that occurs whenever the U. S. takes action in a foreign land.

Ivo Daalder (U. S. permanent representative to NATO) has stated, "Unfortunately, the way Libya has evolved demonstrates that just because you give people the opportunity to decide their own future they don't always decide in the right or best way — in the way that we would have wanted. So the situation in Libya has gone from bad to worse and is horrific in many dimensions. The future doesn't look much brighter" (Robins-Early, 2015). Daalder's perspective is shared by many American statesmen. But it reveals an arrogant, U. S.-centric mental model and a disrespect for local cultures and values. Who are we to decide the "right" or "best" way for any country other than our own? This attitude tends to generate resentment wherever it is applied.

*Systems Thinking Gaffs: the Invasion of Libya*

- Unintended consequences: the U. S. invasion resulted in the rise of militant Islam and chaos in Mali.

- Failure to understand feedback:
  - The strengthening of the Tuaregs disrupted the power balance in Mali.

- o The defeat of Gadaffi's army led a large faction of it to leave Libya and regroup elsewhere.

- Failure to understand root cause systems: again a lack of understanding local Tuareg culture and issues resulted in a tactical victory but a strategic defeat.

- A "Fix that Failed": The Libya invasion was intended to prevent a humanitarian disaster. All it did was transfer the disaster from Libya to Mali.

## VI. Discussion

The foreign policy of the United States can be characterized by major weaknesses in Systems Thinking: poor, unshared, inaccurate mental models; failure to appreciate feedback loops; inability to discover or understand systemic root causes; and misidentification of leverage points.

Mental models such as the U. S. has the right and responsibility to serve as the world's policeman; that democracy is the best form of government (and the U. S. has the right to impose democracy); and that the world shares our philosophical, spiritual, and political values are egocentric and inaccurate. Many of these derive from an inability or unwillingness to study and understand foreign cultures and ways of life.

"The U. S. as the world's Policeman" is a mental model held by some Americans, but widely resented by much of the world. Often, the same oppressed people who wanted the U. S. to intervene to repress a brutal regime end up hating the U.S. more than the oppressive regime. And freedom won by the U. S. military on behalf of an oppressed group is often devalued.

Democracy may not always be the best form of government. In fact, majority rules in a pure democracy and the majority could deny rights to any and all minorities. The United States has a *constitutional* democracy based on the Constitution/Bill of Rights and supported by a legal system, which prevents the denial of rights to any citizen or minority. However, many Middle Easterners do not understand the concept that minorities should enjoy the same rights as the majority, and view a pure democracy as an opportunity gain control. For a democratic government to work, the people must agree that it is the highest law in the land. Democracy may not succeed in a land such as Iraq, in which religious doctrine (as opposed to secular law) is considered the highest law of the land. In Iraq, the ruling Bath party repressed the Shia majority for decades, and the imposed change to democracy provided an opportunity for the Shia to take revenge. Given the history of this region, the imposition of democracy resulted in an unintended consequence to the disadvantage of the Bath party. Systems Thinking analyzes the historical and political ramifications and potential unintended consequences of imposing democracy in Iraq, and concludes that democracy may not be the best form of government for that region. We would certainly resent it if an outside force attempted to impose Sharia law in the Unites States. Why would a non-democratic state view the imposition of democracy any differently?

"People are willing to suffer some collateral damage to achieve philosophical, religious, and political ideals" is another poor mental model. When innocent civilians are killed during air strikes, their friends and families naturally resent it deeply. Some become militant or terrorist. It is therefore unclear if bombing actually reduces or increases the number of anti-American terrorists.

Again, poor mental models often derive from egocentrism: an inability or unwillingness to understand another's viewpoint. The U. S. repeatedly fails to study foreign cultures and values and assumes that all values reflect our own. We must do a better job of developing mental models, exposing them to scrutiny, and understanding the mental models of others.

Linear thinking focuses on tactical, short-term reactions to situations without considering feedback loops. International actions inevitably result in reactions and unintended consequences: the toppling of

Saddam Hussein yielding a power vacuum and the rise of ISIS; the arming of the Mujahedeen strengthening the Taliban and Al Qaeda; the support for the ARPCT in Somalia giving rise to Al Shabaab. Although we can't always predict exactly what those reactions will be, failure to anticipate them is folly. And we have plenty of experience suggesting what reactions are likely to be: large scale invasions will likely yield a resistance; elimination of unappealing leaders will likely yield a power void; arming the enemies of our enemies will often empower a new enemy.

We repeatedly attempt to deal with symptoms instead of underlying systemic root causes which often involve local culture and psychological needs. The appeal of radical, anti-American terrorist organizations is a *symptom*. Understanding the cultural *basis* for this appeal is essential. Many of the geographic areas in which we intervene are characterized by poverty, corruption, oppression, human rights abuses, or years of war and terror. Some local cultures are based on simony, a strong caste system, and nepotism, and youth born into poverty have little hope for a bright future. Appealing to these people with promises of democracy, equality for women, and a western penal code are folly when their basic needs are for food, shelter, healthcare, security, and an economic future. Groups like ISIS promise (and often deliver) these things and, whether totalitarian or not, will always be more appealing to people operating on the lower levels of Maslow's hierarchy than will America's ideological rhetoric.

Alfred McCoy (2016) has a much better grasp of the systemic root cause of the Afghanistan conflict than do our own military and State Department. He writes, "We can continue to fertilize this deadly soil with yet more blood in a brutal war with an uncertain outcome… or we can help renew this ancient, arid land by re-planting the orchards, replenishing the flocks, and rebuilding the farming destroyed in decades of war… until food crops become a viable alternative to opium." His statement applies equally well to many of our military interventions.

Table I shows several iceberg models and the underlying mental models and structures that yield negative events.

172

Table I. U. S. Foreign Policy Iceberg Models

| Iceberg Model Number | Mental Model | Resulting Structure | Pattern | Events |
|---|---|---|---|---|
| 1 | The U. S. is a big, powerful bully that wants to eradicate Islam. | Political, Military, and Social structures that denigrate the U. S. | Increased opposition to U. S. intervention in foreign affairs, repeated anti-U. S. demonstrations and activities, attacks on U. S. embassies and military assets abroad | Anti-U. S. demonstrations and activities, attacks on U. S. embassies, military bases, ships and convoys |
| 2 | Democracy is the best government and the U. S. has the moral obligation to impose it. | The U. S. military and political machine | Repeated intervention in foreign affairs and forced installation of democratic governments | Intervention in countries we don't like and forced installation of democracies. |
| 3 | The U. S. has a moral obligation to be the world's policeman | The U. S. military and political machine | Repeated intervention in foreign affairs | Intervention in foreign affairs |
| 4 | The world shares our philosophical, political, and spiritual values | The U. S. military and political machine | Resentment and support of anti-U. S. terrorism | Anti-American protests, wars, and terrorism |

| | | | | |
|---|---|---|---|---|
| 5 | People are willing to suffer some collateral damage to achieve philosophic al, reli- gious, and political ideals | Military structures resulting in bombings, shootings, and loss of civilian lives | Resentment and support of anti-U. S. terrorism | Anti-American protests, wars, and terrorism |

Figures 4 and 5 depict feedback loops prevalent in our foreign policy.

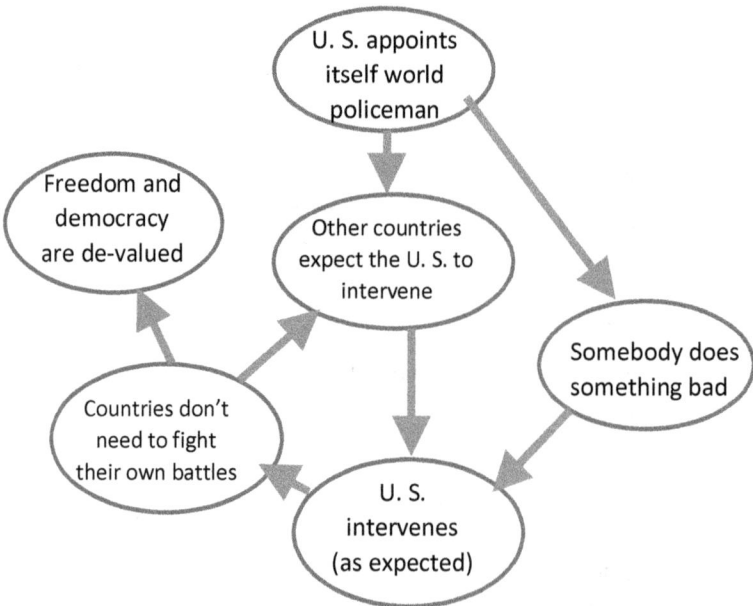

Figure 4. Causal Loop Diagram Showing the Devaluation of Freedom and Democracy due to the U. S. Self-Appointment as the World's Policeman

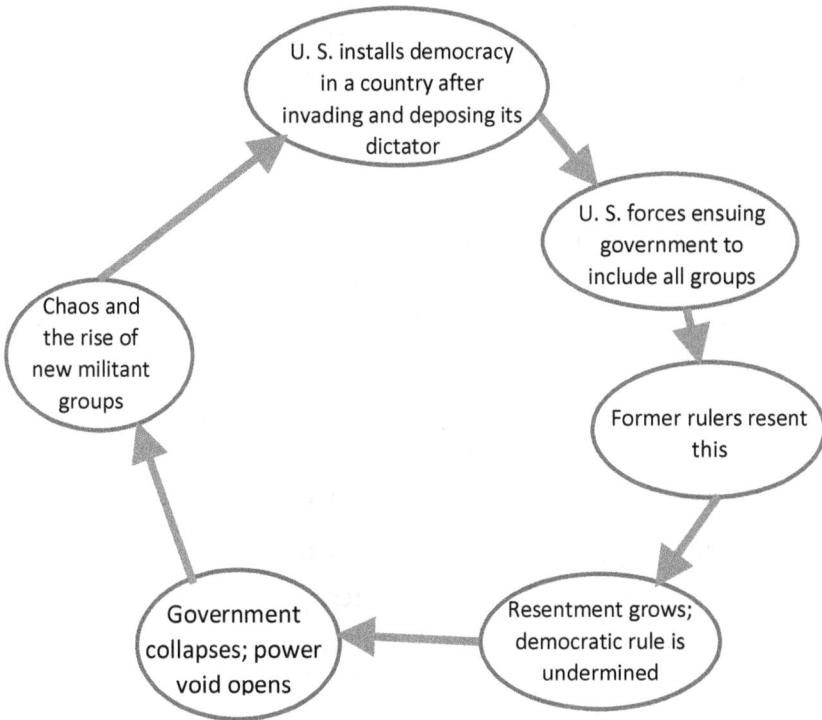

Figure 5. Causal Loop Diagram Showing the Rise of New Militant
Groups as a Result of U. S. Installations of Democracy

VII.    Conclusions and Recommendations

Over the past 40 years, America's foreign policy has not been
particularly successful. Our linear thinking, shoot-from-the hip
approach has yielded negative consequences and could be viewed as
having done more harm than good.  A pertinent quote regarding
Libya comes from NATO representative Ivo Daalder: "Could it have
gone a different way with an outside military intervention? Possibly.
But if we look at the last 25 years, the successes of those foreign
interventions are few and far between……….. Clearly we're learning
a lesson, as we did in Iraq, as we did in Afghanistan, as we're doing
in Syria, as we did in the Balkans, as we did in Somalia and Mali et
cetera. There's a lot to be learned about how one intervenes with a
result that is acceptable and a cost that is equally acceptable. We
haven't found that goldilocks solution yet and we probably never will,
but it doesn't mean we give up and never try or that we take
ownership of these situations and put in troops to stay there for

twenty or thirty years" (Robins-Early, 2015). Unfortunately, it is not clear that we are learning any lesson. The Goldilocks solution requires changing our way of thinking and re-assessing our ultimate objectives.

1. With respect to foreign policy, the U. S. must develop and adopt better mental models, open them to scrutiny, and understand others'. "The U. S. has the responsibility to be the world's policeman," "Democracy is the best form of government," "The enemy of my enemy is my friend," and "The world shares our political, social, economic, and spiritual values" are mental models that have not served us well.

2. We must pay more attention to feedback and unintended consequences developing as a result of actions that we take. Initiatives such as the arming of the Mujahedeen in Afghanistan, the support of ARPCT in Somalia, bombings in Iraq (along with collateral damage), the destruction of poppy fields in Afghanistan, and the forced imposition of democracy in Iraq cause *reactions* from others. Indeed, almost every U. S. action results in pushback from someone. The growth of Al Qaeda, the strengthening of Al Shabaab, the resurgence of the Taliban in Afghanistan, and even the rise of ISIS are unintended consequences of our behavior.

3. We must stop seeking the wrong goal, and work to understand the *Systemic Root Causes* of terrorism in the world. Our knee-jerk reaction typically is to respond militarily. But killing terrorists addresses the symptom, not the disease. Without understanding and addressing the root causes that give rise to terrorism, we will never eliminate it. This involves studying and understanding other peoples' history, culture, values, and economic, political, social, and spiritual systems, and waging socio-economic wars instead of military wars. This is more difficult than sending in troops, but it is essential if we want to solve this problem.

As Alfred McCoy (2016) says, "……. investing even a small portion of all that misspent military funding in rural Afghanistan could produce economic alternatives for the millions of farmers who depend upon the opium crop for employment. Such money could help rebuild that land's ruined orchards, ravaged flocks, wasted seed stocks, and wrecked snowmelt irrigation systems that, before these decades of war, sustained a diverse agriculture. If the international community can continue to nudge the country's dependence on illicit opium down from the current 13% of GDP through such sustained rural development, then perhaps Afghanistan will cease to be the planet's leading narco-state and just maybe that annual cycle can at long last be broken."

Addressing the systemic root cause requires that we decide how the United States should exemplify leadership. Being the military tyrant of the planet has not proven successful. Investing in local systemic structures and mental models that will foster self-sufficiency with respect to food, clothing, shelter, security, and healthcare may prove more fruitful. To amplify Alfred McCoy's suggestion, imagine that instead of spending $3 *trillion* dollars fighting wars in the Middle East (Kiley, 2016), those funds were spent on systemic improvements in housing, hospitals, economic development, and improved agricultural practices. Imagine that as a result of those activities, oppressed people saw a way out of poverty, corruption, and misery, and credited the United States for this. *That* would be leadership.

We hope that our use of Systems Thinking principles to analyze some past foreign policy failures will stimulate the broader use of Systems Thinking by the State Department, military, and intelligence communities when dealing with international issues.

# References

Australian Broadcasting Corporation, "Map Showing Who Has Control of Various Regions of Afghanistan," 21 August 2017, http://www.abc.net.au/news/2017-08-22/map-showing-who-has-control-of-various-regions-of-afghanistan/8831070

Adow, Mohamed Amiin, "Somalia Clashes 'the Worst Since 1991' ", CNN , 21 April 2008.

Afghanistan Language and Culture Program (ALCP), "Ethnic Groups,"___https://larc.sdsu.edu/alcp/resources/afghanistan/people/ethnic-groups/, accessed 22 August 2017.

Agator, Maxime, "Iraq: Overview of Corruption and Anti-Corruption," *Transparency International*, Number 374, https://www.transparency.org/whatwedo/answer/iraq_overview_of_corruption_and_anti_corruption, 10 May 2013.

Almukhtar, Sarah, "The Taliban Still Control Large Parts of Afghanistan and ISIS Has Established a Foothold," New York Times, 6 June 2017.

Asian Wall Street Journal, "Afghanistan's Complex Ethnic Patch-work," *Tehran Times*, 10 March  2011.

Associated Press/CBS News, "Al-Shabab, al Qaeda Alliance a Desperate Bid?" 10 February 2012.

Balaghi, Shiva, "The War on Terror and Middle East Policy Analysis," *Costs of War*, Brown University, 2011.

BBC News, "History: The War in Afghanistan," 2017, http://www.bbc.co.uk/history/the_war_in_afghanistan

Chapman, Steve, "Afghanistan, the 15-year Failure," *Chicago Tribune*, http://www.chicagotribune.com/news/opinion/chapman/ct-afghanistan-war-15years-endless-chapman-perspec-1006-jm-20161005-column.html, October 5, 2016.

Coll, Steve, "Ghost Wars: The Secret History of the CIA, Afghanistan, and Bin Laden, from the Soviet Invasion to September 10, 2001," Penguin Books, London, 2004.

Coll, Steve, "Ghost Wars: How Reagan Armed the Mujahadeen in Afghanistan," Interview with Amy Goodman of *Democracy Now!*, June 10, 2004, https://www.democracynow.org/2004/6/10/ghost_wars_how_reagan_armed_the

Daniels, Jeff, "Defense Secretary James Mattis Admits U. S. 'not winning in Afghanistan'," CNBC, 13 June 2017.

Gane-McCalla, C., How The CIA Helped Create Osama Bin Laden, Newsone, 2011, https://newsone.com/1205745/cia-osama-bin-laden-al-qaeda/

Gouttierre, Thomas, and Matthew S. Baker, "Ethnic Map of Afghanistan," National Geographic Society, 2003.

Hogg, Annabel Lee, "Timeline: Somalia, 1991-2008," **The Atlantic**, December 2008.

Iheduru, Okey C. (29 March 2012). "America and the Rebellion and Coup in Mali", *Business Day*, 29 March 2012, Retrieved 23 November 2012.

James, George, "Somalia's Overthrown Dictator, Mohammed Siad Barre, Is Dead," *The New York Times Obituaries*, http://www.nytimes.com/1995/01/03/obituaries/somalia-s-overthrown-dictator-mohammed-siad-barre-is-dead.html, January 3, 1995.

Kiley, Gillian, "U. S. Spending on Middle East Wars, Homeland Security will Reach $4.79 trillion in 2017," *News from Brown*, Brown University, September 9, 2016.

Kinzer, Stephen, "U. S. Inadvertently Creates a Terrorist Haven in Mali," *Boston Globe*, Sunday, 7/5/2012, p K10.

Maslow, A. H., "A Theory of Human Motivation," *Psychological Review* **50** (4) 370–96, 1943.

*Mazzetti, Mark, "Spy Agencies Say Iraq War Worsens Terrorism Threat,"* *The* New York Times, September 24, 2006.

McCoy, Alfred, "How Opium Defeated America in Afghanistan," *Salon*, February 23, 2016, http://www.salon.com/2016/02/23/washingtons_21st_century_opium_wars_partner/

Mitchell, Andrea, "Trump Afghanistan Strategy Needed 'a Set of Goals,' Panetta says," MSNBC, 22 August 2017, http://www.msnbc.com/msnbc-news/watch/trump-afghanistan-strategy-needed-a-set-of-goals-panetta-says-1029914691813

Monat, J. P., and Gannon, T.F., "What Is Systems Thinking? A Review of Selected Literature Plus Recommendations," Am. J. of Systems Science, 4:2, 2015.

Monat, J. P., and Gannon, T.F., *Using Systems Thinking to Solve Real-World Problems*, College Publications, London, 2017.

Munger, Sean, "The Hunger of War: The Somali Famine of 1991-92," seanmunger.com, 25 February 2015.

National Intelligence Council, "National Intelligence Estimate: Trends in Global Terrorism: Implications for the United States," NIE 2006-02R, April, 2006.

Nordland, Rod, and Fahim Abedjune, "ISIS Captures Tora Bora, Once Bin Laden's Afghan Fortress," New York Times, 14 June 2017.

Qazi, Shereena, and Yarno Ritzen, "Afghanistan: Who Controls What," Aljazeera, 24 Jan 2017.

Robins-Early, Nick, "Was the 2011 Libya Intervention a Mistake?" *The World Post*, http://www.huffingtonpost.com/2015/03/07/libya-intervention-daalder_n_6809756.html, March 7, 2015.

Roggio, Bill, "Afghan Taliban lists 'Percent of Country under the Control of Mujahideen' ," *FDD's Long War Journal*, 28 March 2017.

"Somalia: Reported U. S. Covert Actions 2001-2016," February 22, 2012, Bureau of Investigative Journalism, https://www.thebureauinvestigates.com/drone-war/data/somalia-reported-us-covert-actions-2001-2017.

"The Soviet Invasion of Afghanistan and the U. S. Response, 1978-1980," Office of the Historian, U. S. Department of State, https://

history.state.gov/milestones/1977-1980/soviet-invasion-afghanistan, accessed 23 June 2017.

Sale, Michelle, "Missions Accomplished?" New York Times. April 11, 2003.

Stanford University, "Mapping Militant Organizations: Al Shabaab," http://web.stanford.edu/group/mappingmilitants/cgi-bin/groups/view/61, February 20, 2016.

Stanford University, "Mapping Militant Organizations: Islamic Courts Union," http://web.stanford.edu/group/mappingmilitants/cgi-bin/groups/view/107, March 30, 2016.

United Nations Office on Drugs and Crime, "Corruption Widespread in Afghanistan, UNODC Survey Says," UNODC.org., 19 January 2010.

Washington, Adolphus, "Malian Blues: The Issue of the Tuareg," *Global Risk Insights*, http://globalriskinsights.com/2016/01/malian-blues-the-issue-of-the-tuareg/, January 20, 2016.

Witte, Griff, "Afghanistan War, 2001–2014," Encyclopedia Britannica, 2017, https://www.britannica.com/event/Afghanistan-War

# Applying Systems Thinking to Engineering and Design

Jamie P. Monat and Thomas F. Gannon

Systems Engineering Program
Worcester Polytechnic Institute
Worcester, MA USA

*reprinted from Systems*, Vol. 6, No. 3, 2018, pp. 34-54,
doi:10.3390/systems6030034

**Abstract:** The application of Systems Thinking principles to Systems Engineering is synergistic, resulting in superior systems, products, and designs. Yet there is little practical information available in the literature that describes how this can be done. In this paper, we analyze 12 major Systems Engineering failures involving bridges, aircraft, submarines, water supplies, automobiles, skyscrapers, and corporations and recommend Systems Thinking principles, tools, and procedures that should be applied during the first few steps of the System Engineering design process to avoid such catastrophic Systems Engineering failures in the future.

**Keywords:** Systems Thinking, Systems Engineering, design

---

## 1. Introduction

### 1.1. Systems Thinking and Systems Engineering

Systems Thinking and Systems Engineering are not the same.

Systems Thinking has been characterized as a perspective, a language, and a set of tools [1]. It is a holistic perspective that acknowledges that the relationships among system components and between the components and the environment are as important (in terms of system behavior) as the components themselves. It is a language of feedback loops, emergent properties, complexity, hierarchies, self-organization, dynamics, and unintended consequences. Systems Thinking tools include the Iceberg model which posits that in systems, repeated events and patterns (which are observable) are caused by structure (stocks, flows, and feedback loops) which are, in turn, caused by underlying forces such as mental models, gravity, and electromagnetism. Additional Systems Thinking tools include causal loop diagrams, behavior-over-time plots, stock-and-flow diagrams, systemic root cause analysis, dynamic modeling tools, and archetypes. A more comprehensive explanation of Systems Thinking is provided by Monat and Gannon [1].

Systems Engineering is an interdisciplinary approach and means to enable the development of successful systems. It focuses on defining customer needs and required functionality early in the development cycle, documenting requirements, and then proceeding with design, synthesis, validation, deployment, maintenance, evolution and eventual disposal of a system. Systems Engineering integrates a wide range of engineering disciplines into a team effort, which uses a structured development process that proceeds from an initial concept to production and operation of a system. It takes into account both the business and technical needs of all customers with the goal of providing a quality product that meets the needs of all users. A more comprehensive description of Systems Engineering and the structured processes used by Systems Engineers to develop systems is provided by Kossiakoff *et al* [2].

Table 1 shows the Systems Thinking concepts that are applicable to Systems Engineering and Design.

**Table 1.** Systems Thinking Concepts Applicable to Systems Engineering and Design.

| Systems Thinking Principle | Applicable to Systems Engineering and Design? |
| --- | --- |
| Holistic Perspective/Proper Definition of System Boundaries | Yes |
| Focus on Relationships | Yes |
| Sensitivity to Feedback | Yes |
| Awareness that Events and Patterns are caused by Underlying Structures and Forces | To a Degree |
| Dynamic Modeling | Sometimes |
| Systemic Root Cause Analysis | Yes, when troubleshooting |
| Sensitivity to Emergent Properties | Yes |
| Sensitivity to Unintended Consequences | Yes |

Systems Thinking and Systems Engineering are related and synergistic, and the application of Systems Thinking Principles to Systems Engineering can result in superior systems. Yet there seems to be a dearth of practical information describing how this can be done.

Lawson [3] discusses the natural coupling of thinking and engineering systemic structures, and gets into the specifics of applying Systems Thinking principles and tools (such as a System Coupling Diagram) to Systems Engineering. He couples thinking with acting via OODA (Observe-Orient-Decide-Act) and PDCA (Plan-Do-Check-Act) loops [4, 5]. The PDCA loop seems especially relevant to systems engineering. Lawson provides a good example of the application of Systems Thinking principles to the development of automatic train control in Sweden in the 1970s. Kasser and Mackley [6] do a very nice job describing various Systems Thinking Perspectives (Operational, Functional, Big picture, Structural, Generic, Continuum, Temporal, Quantitative, Scientific) and applying them (by way of example) to the Royal Air Force (RAF) Battle of Britain Air Defence System (RAFBADS) by asking "Who, what, why, where, when, and how?" for each perspective. The Systems Thinking perspectives represent a good tool; however, more tools and specifics would be useful in inculcating Systems Thinking into Engineering and Design. Godfrey [7] suggests that Systems Thinking is essential for Systems Engineers and that Systems Thinking provides a key underpinning for Systems Engineering. Godfrey's work supports Kasser's "Who, what, why, where, when, and how?" approach but focuses on explaining Systems Thinking as opposed to its application to and integration with Systems Engineering. Pyster *et al.* [8] obliquely relate Systems Thinking to Systems Engineering via the application of patterns and archetypes. The INCOSE Systems Engineering Handbook [9] describes Systems Thinking and lists several Systems Thinking principles that Systems Engineers are likely to encounter. It does not, however, prescribe a methodology or integrate Systems Thinking tools with Systems Engineering. Burge [10] extols the virtues of Systems Thinking in engineering design and points out that it may be used in several different ways: to gain understanding of a complex situation, to gain sufficient understanding to make predictions of future system behavior, to solve a problem, or to create a new or modified system; however, he does not explain *how* this is to be done. Frank [11] explores the cognitive competencies (several of which involve Systems Thinking) of successful Systems Engineers; but he does not discuss how Systems Thinking principles or tools should be applied to Systems Engineering.

There are also instructive examples of Systems Engineering failures due to inattention to Systems Thinking principles.

*1.2. Systems Engineering Failures due to Lack of Systems Thinking*

There have been many Systems Engineering and design failures due to a lack of applying Systems Thinking. In this section, we describe a few of the more infamous cases.

**Figure 1.** The Tacoma Narrows Bridge Collapse

*Galloping Gertie.* The infamous Tacoma Narrows Bridge linked Tacoma, Washington to the Kitsap Peninsula from 1938-1940. The bridge spanned the Tacoma Narrows, which was known for high winds. Those high winds caused both an up-and-down roller-coaster like oscillation and aeroelastic *torsional flutter*: an oscillatory twisting of the bridge deck [12]. This latter phenomenon comprised a destructive reinforcing feedback mechanism when winds blew horizontally across the bridge deck (similar to the sound-inducing vibration created by blowing across a taught blade of grass.) When the windward edge of the deck flexed slightly up or down (in this case due to a support cable failure,) more of the deck's horizontally-projected surface area was exposed to the wind, yielding higher wind forces, which twisted the deck even more. The twist increased until the deck's torsional restoring force returned it to horizontal *and beyond*, and the process repeated, yielding a sinusoidal torsional oscillation. (a good video of this phenomenon and the subsequent collapse may be found at https://www.youtube.com/watch?v=j-zczJXSxnw.) Vortex shedding downwind of the deck exacerbated the oscillations. The torsional oscillations were usually adequately damped by the stiffness of the bridge deck, but with the unfortunate combination of high wind speed and a snapped support cable (occurring on 7 November 1940), the damping declined to near zero. At this point, the torsional oscillations grew ever larger, until the bridge collapsed (Figure 1).

*Specific System Thinking Oversight:* Failure to bound the system properly and to adequately evaluate system component interactions with the environment under all conditions, especially with respect to feedback. The original bridge design had called for 25-foot high trusses under the deck, which would have stiffened the deck sufficiently to prevent the collapse. But to save money, that design was supplanted with a new design that replaced the trusses with 8-foot high steel plates, saving $3 million, but significantly reducing torsional resistance. The bridge designers might be excused had there been no prior experience with wind-induced bridge failure and

185

torsional flutter. But these phenomena had both been observed in the late 19th century [13]; in this case they were simply forgotten or ignored [14].

Figure 2. London's Millennium Footbridge

*Millennium Bridge, London.* The Millennium Bridge (Figure 2) is a pedestrian bridge crossing the Thames River in London. It was built from 1998-2000 and opened to pedestrian traffic in 2000. The bridge design (the result of a competition) was novel, with a shallow profile and the supporting cables *below* the deck [15]. Shortly after opening, pedestrians noticed a peculiar lateral swaying motion that caused them to naturally adjust their stride in synchronization with the sway [16]. This synchronization amplified the sway as the bridge was driven at its natural *lateral* resonance frequency in a reinforcing feedback loop (the sway caused pedestrians to march in step with the sway; the marching in step amplified the sway.) Vertical resonance caused by wind or by soldiers marching in step was well-known to bridge designers, but lateral resonance had not been anticipated in this new, modern design [17]. Eighty-nine vertical and horizontal dampers were installed on the bridge to correct this problem, at a cost of ~$7 million.

**Specific System Thinking Oversight:** Failure to adequately evaluate system component interactions under all use conditions, especially with respect to feedback. Despite thorough engineering analysis prior to construction (including modeling and wind tunnel testing to assess both lateral and torsional impacts of *vertical* pedestrian excitation) the analyses did not consider synchronous *lateral* excitation. There was reference to this phenomenon in the literature [18, 19], but building codes did not mention it. Subsequent to the bridge's construction, BD 37/01 (the British Standard on bridge live loading) was revised to include a section on synchronous lateral excitation.

*The Lockheed L-188 Electra Turboprop Airplane.* The L-188 Electra was a 4-engine turboprop airplane developed by Lockheed in the late 1950s. The

186

plane suffered 2 fatal crashes in 1959 (Braniff 542) and 1960 (Northwest 710), killing 97 passengers and crew. In each of these crashes, excessive wing flutter and vibration led to the wings shearing off the fuselage. Analysis and wind tunnel tests revealed a fatal reinforcing mechanical feedback loop involving the plane's engine mounts. It was determined that the engine mounts permitted small oscillations of the engine on the wing. These engine oscillations caused the wing to flutter which, in turn, caused the engine to oscillate more. The increased engine oscillation increased the wing flutter. This reinforcing feedback loop caused the wings to vibrate at their natural resonant frequency, eventually breaking away from the fuselage. In this situation, mechanical feedback was not considered in either the design stage or the engineering test stage. It is interesting to note that after the crashes the engine oscillations were subsequently reproduced in wind-tunnel testing of scale models.

*Specific System Thinking Oversight*: Failure to properly evaluate system component interactions under all use conditions, especially with respect to feedback.


*The Water of Ayolé*. The Water of Ayolé demonstrates that few engineering/technical issues are exclusively technical; most involve support infrastructures, people, the environment, economics, and other factors. Ayolé is a small rural village in the West African country of Togo. In the 1970s-80s the water source for the village was the Amou River, which happened to be infested with the guinea worm *Dracunculus medinensis,* a parasite that infects a human host and causes excruciating pain. To address this issue, government and international aid organizations dug and installed wells in the village, which worked well for several years. But as the wells broke down (due to normal wear and tear) no spare parts were available, no technical expertise was available to fix or maintain the pumps, and no money was available to pay for repairs. After 3 years, the people of Ayolé were back to using the contaminated water from the river. The government engineers had interpreted this as a purely technical/engineering problem, when in fact it was much broader. To their credit, the local Togolese extension agents applied Systems Thinking to address the larger systemic issues. They trained some of the villagers in well maintenance and repair; they established a repair parts supply chain via the local Togo hardware store; and the women of the village structured an agricultural product production and sales system to generate money to pay for the parts. What was thought to be a simple engineering problem turned out to be an engineering/socio-economic/ logistics/ psychological problem.

*Specific System Thinking Oversight*: The assumption that real-world engineering problems can be solved by purely technical means. Many involve sociological, psychological, economic, legal, maintenance, support, and "soft" human issues.


*Stow Center School Aquarium.* A situation similar to that of Water of Ayolé occurred on a much smaller scale in 2003 at the Center School in Stow, MA, USA. A local university had donated a salt-water aquarium to the school, complete with filtration equipment, temperature regulation, and fish. The students loved the aquarium and were fascinated by its horseshoe crabs and other exotic-looking sea creatures, enjoying it for several years. But then, components began to wear out and needed replacement, the system needed cleaning and maintenance, and the creatures required regular feeding and assessment. No one had been assigned the responsibility for system maintenance, and teachers (who often felt underpaid) did not feel that maintenance was their responsibility. Eventually, the beautiful, free aquarium was disassembled and scrapped.

**Specific System Thinking Oversight**: Like the Water of Ayole, there were ancillary support and maintenance issues associated with the aquarium; issues rarely have purely technical solutions. The well-intentioned donors did not realize that this "gift" imposed a burden on the recipients.


*The Russian K-141 Kursk Submarine Disaster.* On August 12, 2000, the Russian Kursk nuclear submarine exploded and sank off the coast of Russia in the Barents Sea; all 118 crewmen were lost. The explosion was traced to a leak of hydrogen peroxide ($H_2O_2$) from one of the ship's torpedoes; the peroxide reacted explosively with copper or brass that was present in the torpedo tube.

Both hydrogen peroxide alone and the combination of kerosene and hydrogen peroxide had been used as propellants for rockets and torpedoes since the 1930s [20]. It was well-known that hydrogen peroxide reacts violently in the presence of a silver, copper, or brass catalyst, and previous near-disasters with $H_2O_2$-powered torpedoes had been well-documented. Because of this risk, hydrogen peroxide/kerosene propellants have been banned by the British and other navies and replaced by newer, safer combinations such as Otto Fuel and Hydroxyl Ammonium Perchlorate [21, 22]. When the peroxide leaked from the torpedo and contacted the catalyst on the Kursk, it was converted to oxygen and water vapor, increasing in volume by a factor of 5,000. This sudden pressure surge caused a subsequent explosion (equivalent to 500 pounds of TNT) in a nearby kerosene tank and blew a hole in the submarine's hull. It is thought that the sub then sank within minutes. A few minutes later (and while on the ocean floor), the heat from the kerosene explosion resulted in the detonation of several nearby torpedoes

[23]. The explosions blew an immense hole in the sub, and most of the sub's compartments flooded. Subsequent confusion and lack of transparency by the Russians prevented a rescue of the trapped sailors.

*Specific System Thinking Oversight*: Failure to identify both system component <u>planned</u> inter-relationships and <u>unplanned</u> inter-relationships. The knowledge of the catalytic reaction of hydrogen peroxide with metals (with potentially devastating consequences) should have prevented the co-location of any hydrogen peroxide containers with metal catalysts.

**Figure 3.** The Vdara Hotel

*The Vdara Hotel, Las Vegas.* Las Vegas's Vdara hotel (Figure 3) was designed by Rafael Viñoly and built in 2008. The hotel's curved façade focuses the sun's rays like a parabolic reflector, heating the pool area at its base to over $135°$ F—an unintended consequence [24, 25]. Employees and visitors refer to the effect as the "Death Ray." Non-reflective film has been applied to the hotel's highly-reflective windows and large umbrellas have been placed around the pool area, but the deck still gets hot. The area gets so hot, in fact, that guests have reported burning skin and singed hair within minutes of lying near the focus [26].

*Specific System Thinking Oversight*: failure to include relevant environmental components such as the sun and its interaction with other system components.

189

**Figure 4.** 20 Fenchurch Street, London

*20 Fenchurch Street, London.* In London's financial district, 20 Fenchurch Street (Figure 4) rises 34 stories from the street. Also designed by Rafael Viñoly and completed in 2014, the parabolic shape of the highly reflective upper stories focuses sunlight onto a small area at street level for several hours each day, resulting in storefront temperatures exceeding 200 °F. An automobile was partially melted [27, 28] and a reporter fried an egg on the sidewalk [29]. Local businesses were negatively affected by the intense heat. The thermal behavior caused locals to nickname the building the "Walkie-Scorchie" [30] and the "Fryscraper" [31]. Louvers, shades, and non-reflective glass have been considered as remediation measures.

*Specific System Thinking Oversight*: One would think that Viñoly would have learned his system definition lesson after the Vdara debacle several years earlier. Such was not the case. Viñoly repeated the error of not including all relevant environmental components and their interrelationships in his design; in this case the interaction of the sun with the building.

*Toyota Gas Pedal/Floor Mat Entrapment.* In 2007-2010, several accidents and fiery deaths were attributed to "sudden acceleration" in Toyota vehicles. Camrys, Avalons, Highlanders, Matrixes, Priuses, Venzas, Tacomas, and Tundras, and Lexus ES350s, IS250s, and IS350s were affected [32]. The problem was at first denied, and then traced to unintended interactions between the vehicles' floor mats and accelerator pedals [33]. It seems that in certain circumstances, the gas pedal would "bond" to the floor mat resulting

190

in unintended acceleration, inability to slow or stop the vehicle, and consequent accidents. The company recalled 4.2 million vehicles for floor mat replacements and potential gas pedal redesign.

*Specific System Thinking Oversight*: This is an interesting case (similar to the Kursk) in which the system components were properly identified, but the unintended interactions among components were not.

*Biodegradable German Car Wiring Insulation.* The Green Party in Germany passed a law in the early 1990's that required a certain percentage of the parts in an automobile to be bio-degradable, and the EU followed suit in the mid-1990's. Mercedes-Benz decided to rely on bio-degradable wiring insulation to meet those requirements. Unfortunately, a bio-degradable wiring system that is exposed to the environment will eventually decompose into a mass of short-circuiting copper wires [34].

*Specific System Thinking Oversight*: This situation is similar to the Galloping Gertie fiasco in which environmental factors (weather) were not adequately considered.

*The Bhopal Disaster.* A Union Carbide pesticide plant in Bhopal, India resulted in 2,259 immediate deaths and some 11,000 delayed deaths following an accident in 1984. A highly toxic material called Methyl isocyanate, used in the making of pesticides, became contaminated with water, which caused an exothermic reaction that increased the temperature inside a tank well beyond its capacity. An automated emergency release system vented the extra pressure and a large volume of gasses escaped and spread to the surrounding town, killing over 13,000. If one were to define the "system" as just the pesticide plant, and consider that system in isolation, then the safety sub-system worked pretty well, relieving the pressure and preventing an explosion. However, when including the surrounding town and people in the definition of the "suprasystem" [35], it was an utter disaster. (A "suprasystem" is a larger system that integrates several smaller systems. In our usage here, it means the system proper plus the environment, system users, system controllers and maintainers, communications to and from the system, and power to the system.) Additionally, if the vented gasses had been lighter than air, they might have dispersed without much harm. Unfortunately, those gasses were heavier than air, and seeped into the nearby city of Bhopal at ground level. That leak caused over 550,000 injuries.

*Specific System Thinking Oversight*: This situation is similar to the Galloping Gertie fiasco in which the interactions of the system components with each other and with the environment (people in the surrounding town, weight of gases) were not considered adequately.

*The Microsoft Zune.* In response to Apple's fabulously successful iPod (released in 2001,) Microsoft released its own portable music player, the Zune, in 2006. The Zune did not have the iPod's aesthetic appeal or "cool" factor [36]. Perhaps more significantly, Microsoft did not appreciate that the Zune (and all personal media players) are part of a *User Experience System,* and that to be successful, all components of that system must be addressed. Steve Jobs and Apple were fabulous at this. Don Norman says, "It is *not* about the iPod; it is about the **system**. Apple was the first company to license music for downloading. It provides a simple, easy to understand pricing scheme. It has a first-class website that is not only easy to use but fun as well. The purchase, downloading the song to the computer and thence to the iPod are all handled well and effortlessly. And the iPod is indeed well designed, well thought out, a pleasure to look at, to touch and hold, and to use. There are other excellent music players. No one seems to understand the systems thinking that has made Apple so successful" [37]. Although Microsoft eventually attempted to develop a computer interface, a music subscription service ("Zune Music Pass"), and download capabilities ("MSN Music" and "Zune Marketplace"), those efforts were too little, too late, and the *User Experience System* was clearly not Microsoft's focus. The Zune failed commercially and was discontinued in 2011. It should be noted that other portable music device manufacturers (Sony, Diamond, Tascam) also missed the systems thinking aspects of product design: engineering the product is not the same as engineering the *User Experience System.*

**Specific Systems Thinking Oversight**: Failure to understand that many products are not really stand-alone devices, but instead are merely one component of a *User Experience System.*


These failures demonstrate the need to apply Systems Thinking principles (especially appropriate "system" definition, system boundaries, identification of all relevant system *and environmental* components and relationships, and consideration of feedback mechanisms) to Systems Engineering and design. They are summarized in Table 2.

| Problem | Systems Thinking Issue |
|---|---|
| Galloping Gertie | Failure to adequately address planned and unplanned interactions between system components and the environment. |
| Millenium Bridge, London | Failure to adequately address planned and unplanned interactions among system components themselves and between system components and the environment. |
| Lockheed L-188 Electra Turboprop Airplane | Failure to adequately address planned and unplanned interactions among system components themselves and between system components and the environment. |
| Water of Ayolé | Failure to bound the system properly; specifically, to understand that most complex problems cannot be solved by purely technological means; they often involve organizational, political, economic, environmental, ethical, and sociological components. |
| Stow Center School Aquarium | Failure to bound the system properly; specifically, to understand that most complex problems cannot be solved by purely technological means; they often involve organizational, political, economic, environmental, ethical, and sociological components. |
| Russian K-141 Kursk Submarine | Failure to adequately address planned and unplanned interactions among system components themselves and between system components and the environment. |
| Vdara Hotel | Failure to identify relevant environmental factors. |
| 20 Fenchurch Street, London | Failure to identify relevant environmental factors. |
| Toyota Gas Pedal | Failure to adequately address planned and unplanned interactions among system components themselves and between system components and the environment. |
| Biodegradable German Car Wiring Insulation | Failure to identify relevant environmental factors. |
| Bhopal, India | Failure to identify relevant environmental factors; specifically, interactions between the system and |

| | |
|---|---|
| Microsoft Zune | Failure to recognize that many products are actually components of a *User Experience System* |

neighboring people and between the system and environmental gases.

---

It is interesting that the Systems Thinking failures explaining these 12 problems fall into just 4 categories:

a. Failure to identify relevant environmental factors such as wind, insolation, rain, temperature

b. Failure to understand that most complex problems cannot be solved by purely technological means; they often involve organizational, political, economic, environmental, ethical, and sociological components

c. Failure to adequately address both planned and unplanned interactions among the system components themselves and between system components and the environment.

d. Failure to recognize that many products are actually components of a *User Experience System*

## 2. How to Engineer and Design Systems to Avoid Similar Issues

*2.1. Identification of relevant environmental factors such as wind, insolation, rain, temperature*

Failure to identify a system's relevant environmental factors can result in disaster. In the Biodegradable German Car Wiring Insulation debacle, failure to consider normal weather impacts on the insulation caused catastrophic failures. In the Vdara hotel situation (and the 20 Fenchurch Street example), failure to consider the interaction of the building with the sun yielded poor results.

It is sometimes difficult to decide what is part of a system versus what is part of the system's environment. Kossiakoff [2] suggests 4 criteria to decide which elements are system versus environmental components:

1. Developmental Control: the ability of the system designer to control the component

2. Operational Control: the ability of the system operator to control the component

3. Functional Allocation: the ability of the system designer/operator to assign functions to the component

4. Unity of Purpose: the degree to which the component is dedicated to the system's successful performance

Components for which these 4 criteria score low are defined as environmental (not system) components. Kossiakoff suggests depicting system boundaries using a *Context Diagram* (see Figure 5, adapted from Johns Hopkins [38]).

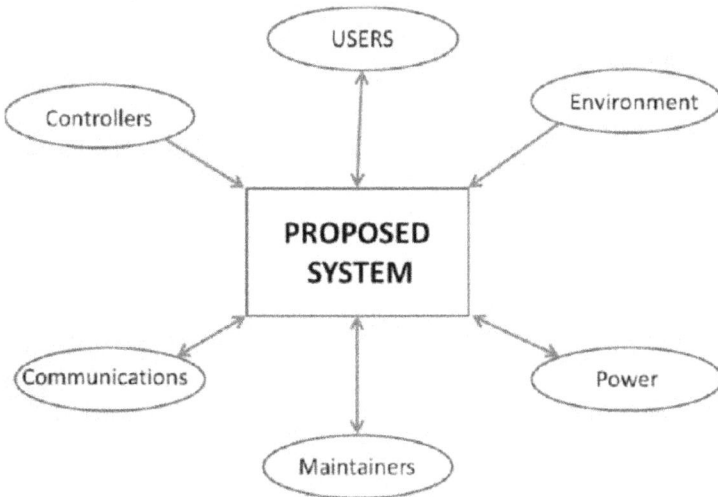

**Figure 5.** System Context Diagram [38]

The context diagram notes additional entities (environment, users, maintainers, etc.) that must be considered in system design. For present purposes, we will focus on the environment. To identify environmental factors that may impact system performance, more detail and specificity are required for both system and environmental components. An excellent tool for this is the *System Breakdown Structure* or SBS.

A SBS is a hierarchical pictogram showing the systems, environment's, and user's components (see Figure 6). (Subsequently, the interrelationships among these components will be described; but first the components themselves must be identified.) Additional sublevels may be added to whatever degree is necessary. For example, Sub-System 1 may comprise several sub-sub-systems. Often, non-Systems Thinkers focus only on the tangible physical components of a system; this can lead to problems as evidenced by Table II. Note that Figure 6 represents a good starting point for a SBS; however, it should be tailored for each specific system design. For example, designers of offshore oil rigs would need to include more environmental factors related to seas, the ocean floor, and underwater life. The top levels of the SBS correspond to the elements depicted in Figure 5; however, the SBS contains much more detail and a level of specificity that is actionable.

195

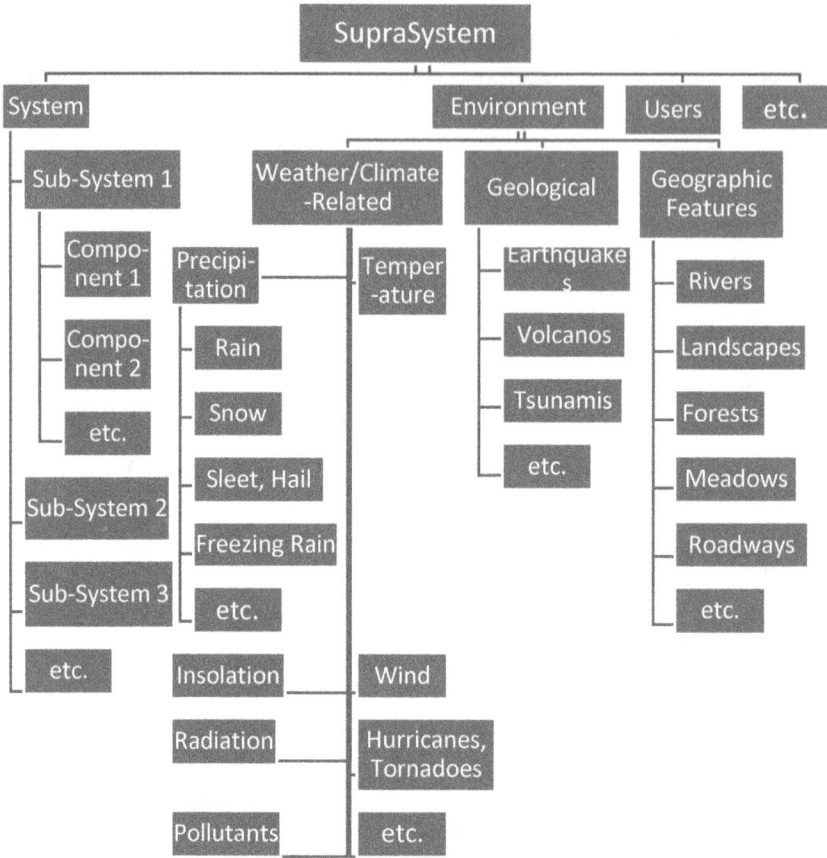

**Figure 6.** System Breakdown Structure (SBS)

*2.2. Bounding the System Properly with respect to understanding that most complex problems cannot be solved by purely technological means; they often involve organizational, political, economic, environmental, ethical, and sociological components*

In the Water of Ayolé example, a narrow perspective of the problem as only an engineering issue led to failure. This was repeated in the Stow Center School aquarium situation. It is a common oversight. Few technical issues do not have human, environmental, economic, sociological, or emotional issues associated with them, yet many engineers are uncomfortable dealing with these "softer" aspects of engineering.

In Ayolé, the government assumed that installing pumps would solve the water problem. Yet 3 years after installation, the pumps were no longer functional. The following sub-systems needed to be created and installed to fully solve the problem:

i. A water and pump operation training and education system

ii. A cultural sensitivity system

iii. A pump maintenance and repair system

iv. A supply chain for repair parts

v. A money-generation and management system, including a farming sub-system

vi. A village/social organization to appropriately divide the labor and decision-making

It is not hard to generalize from these examples when designing or engineering a new or modified system:

i. Will the new system or systemic change require any cultural adjustment? If so, cultural sensitivity and training will be required.

ii. Will the new system require training and education in its use, benefits, and maintenance?

iii. Will a cleaning, maintenance, and repair system need to be established?

iv. Will a repair parts supply chain need to be established?

v. Is there a means to pay for repair, maintenance, legal issues, decommissioning?

*2.3. Adequately addressing both planned and unplanned interactions among the system components themselves and between system components and the environment.*

Several of the examples listed (Kursk, Galloping Gertie, Lockheed Electra, Millennium Bridge, Toyota Gas Pedal, Biodegradable German Car Wiring, Bhopal) resulted from failure to identify potential interactions among system components, or between system components and the environment. Several tools are available to minimize the probability of overlooking these relationships:

2.3.1. System Interrelationship Matrix.

One of the best tools for identifying planned and unplanned relationships among system components and among system and environment components is the System Interrelationship Matrix or SIM. One constructs the SIM by listing all system and environmental components on both axes of a 2-dimensional matrix, as shown in Figure 7. Then one places an X in every cell representing an interaction between the components. One may add detail by noting (in addition to the X) the type of interaction: for example, command-control, mechanical, chemical, emotional/psychological, organizational, frictional. One may make a single matrix for the entire suprasystem, or one may develop smaller, more tractable SIMs for sub-levels or components, as shown in Figure 8 for automobile components.

| | Engine | Brakes | Trans-mission | Suspen-sion | Ignition System | Fuel System | Driver | Passen-gers |
|---|---|---|---|---|---|---|---|---|
| Engine | - | | X | | X | X | | |
| Brakes | | - | | X | | | X | |
| Trans-misson | X | | - | X | | | X | |
| Suspension | | X | X | - | | | | |
| Ignition System | X | | | | - | | | |
| Fuel System | X | | | | | - | | |
| Driver | | X | X | | | | - | X |
| Passengers | | | | | | | X | - |

Figure 7. SIM for the Highest Level of an Automobile

| Suspension Sub-System | Wheels | Tires | Springs | Shocks and Struts | Linkages | Bushings, Bearings, Joints | Steering System |
|---|---|---|---|---|---|---|---|
| Wheels | - | X | X | X | X | X | X |
| Tires | X | - | | | | | |
| Springs | X | | - | X | X | X | X |
| Shocks and Struts | X | X | | - | X | X | X |
| Linkages | X | X | X | | - | | |
| Bushings, Bearings, Joints | X | X | X | | | - | X |
| Steering System | X | X | X | | | X | - |

Figure 8. SIM for the Components of an Automobile Suspension

System Interrelationship matrices have the great benefit of comprehensiveness; however, they can become unwieldy.

2.3.2. Stock-and-Flow Diagrams.

Stock and Flow diagrams depict the interrelationships among system components, and between systems and their environments, from a control volume perspective. Monat and Gannon [39] state, "In systems, some quantities are stored while others flow. These may be real physical quantities such as dollars, volume of water, number of customers, or number of cabbages in a field. They may also be non-physical quantities such as love, anger, greed, or other emotions. Stores or accumulations of these items are called "stocks". Stocks increase or decrease as quantities flow into or out of them. Like causal loop diagrams, stock-and-flow diagrams are helpful in understanding systemic behavior." An example of a stock-and-flow diagram is shown in Figure 9.

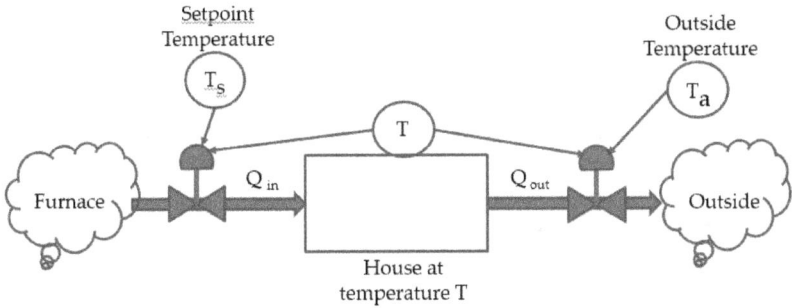

**Figure 9.** Stock-and-Flow Diagram for Domestic Heating System Control

A weakness of stock-and-flow diagrams is that it is hard to determine if one has been comprehensive in showing *all* stocks and flows.

2.3.3. Causal Loop Diagrams.

Causal Loop Diagrams (CLDs) are another tool that may be used to show cause-and-effect relationships among system and environmental components. They are especially helpful in depicting feedback processes, which are present in most systems. Some feedback loops (such as the stabilizing feedback mechanism of a Proportional-Integral-Differential (PID) controller or the reinforcing feedback mechanism of compound interest) are obvious, whereas some (financial bailouts, trade tariffs) are more subtle.

An extremely simple version of a domestic heating system CLD is shown in Figure 10.

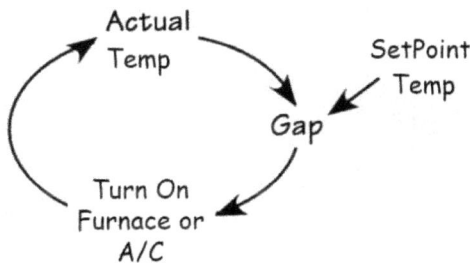

**Figure 10.** Domestic Heating System Causal Loop Diagram

CLDs can become complicated as various cause-and-effect relationships are identified and depicted. A more complicated CLD (related to the Water of Ayolé example) is shown in Figure 11. This CLD highlights the interactions among the water system and the sociological impacts of the system.

199

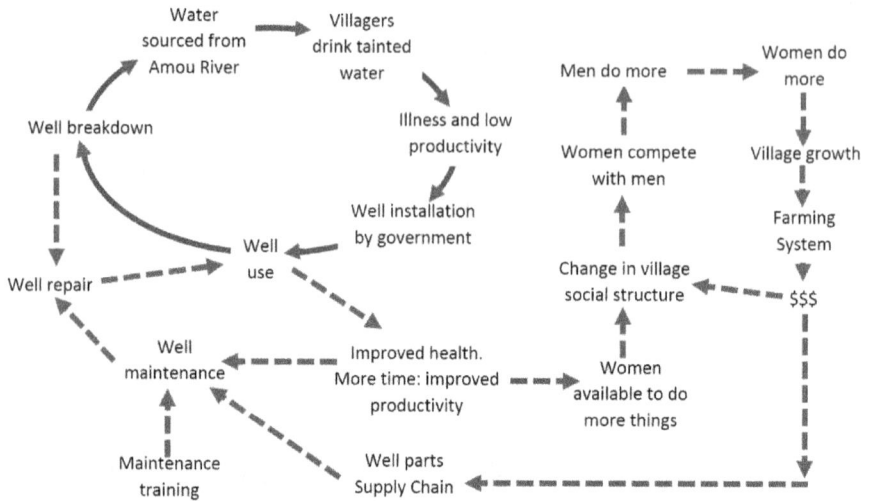

**Figure 11.** CLD for the Water of Ayolé Example. The solid arrows represent the initial effort; the dashed arrows represent the additional structure after the final effort.

As for stock-and-flow diagrams, CLDs are useful, but it is hard to determine if one has been comprehensive in capturing all interrelationships. CLDs are explained in greater detail in *The Systems Thinker* [40] and in Kim [41].

2.3.4. N² Diagrams.

An N² chart or N² diagram is an N x N matrix designed to show interfaces based on system function. Various system functions are plotted on the matrix diagonal; inputs are shown vertically (up or down) and outputs are shown horizontally (left or right). A generic N² diagram is shown in Figure 12 while a specific one is shown in 13.

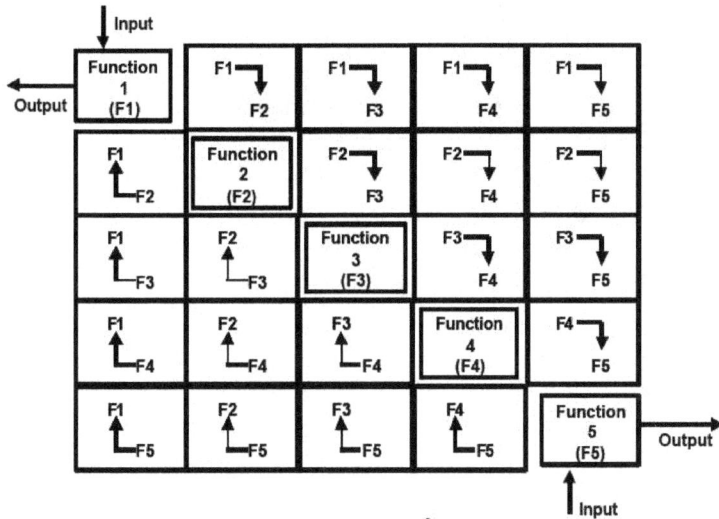

**Figure 12.** Generic N² Diagram from FAA [42]

Feedback loops may be shown as closed circles and critical functions are identified as cells in which several circles intersect. N² diagrams are useful constructs; however, it is not clear that they are comprehensive and identify *all* interactions. For example, they show only interfaces between functions; some system and environment components are not functions and therefore those interrelationships may be missed.

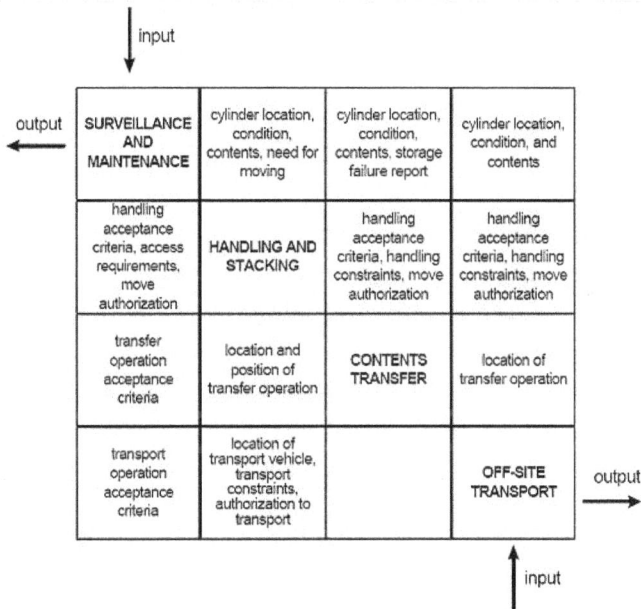

**Figure 13.** Specific N² Diagram for "Functional Flow of Operations" for UF₆ Cylinder Management, from Bechtel Jacobs Company [43]

201

## 2.3.5. SV-3 System-System Matrix.

The SV-3 Systems-Systems Matrix is a DoD construct designed to summarize system resource interactions, manage interfaces, and compare interoperability characteristics of solution options [44]. The DoD website states, "The SV-3 is generally presented as a matrix, where the Systems resources are listed in the rows and columns of the matrix, and each cell indicates an interaction between resources if one exists. Many types of interaction information can be presented in the cells of a SV-3. The resource interactions can be represented using different symbols and/or color coding that depicts different interaction characteristics, for example:

a.     Status (e.g., existing, planned, potential, de-activated).

b.     Key interfaces.

c.     Category (e.g., command and control, intelligence, personnel, logistics).

d.     Classification-level (e.g., Restricted, Confidential, Secret, Top Secret).

e.     Communication means (e.g., Rim Loop Interface, Scalable Loop Interface)."

An example of an SV-3 matrix, based upon the development of the Mobil SpeedPass, is shown in Figure 14.

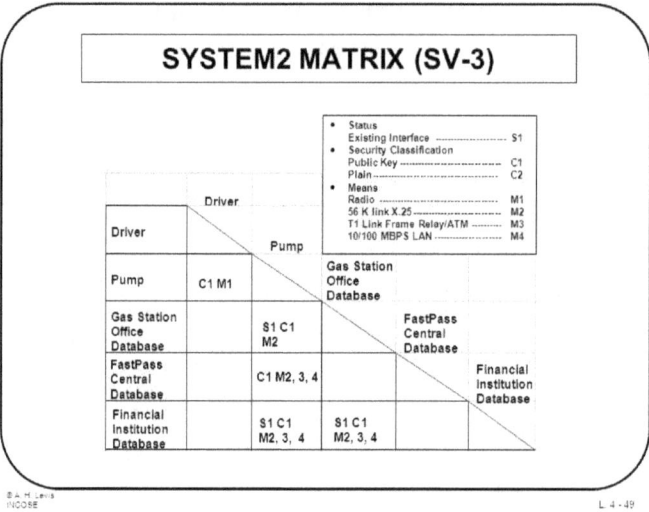

**Figure 14.** Mobil SpeedPass SV-3 Matrix [45]

One weakness of the SV-3 is that it depicts interfaces only among systems and subsystems, not among components. Therefore, some key interrelationships may be missed.

*2.4. Recognizing that many products are part of a larger User Experience System.*

Steve Jobs, founder of Apple Computer was a great Systems Thinker. He recognized that most consumer products are not stand-alone products, but part of experiential *consumer use systems*. While competitors like Sony, Tascam, Microsoft, Diamond) structured their companies around their products (such as portable music players), Jobs structured Apple around the consumer's *experience*, of which the product (in this case the iPod) was just one component. The device itself is just one element of listening to music; other elements include the means of acquiring the music, the user's activities while listening, the environment while listening, and the prestige/coolness factor that may be associated with both the product and the listening experience [37] (see Figures 15 and 16 ).

**Figure 15.** Sony, Tascam, Microsoft, Diamond Structure [39]

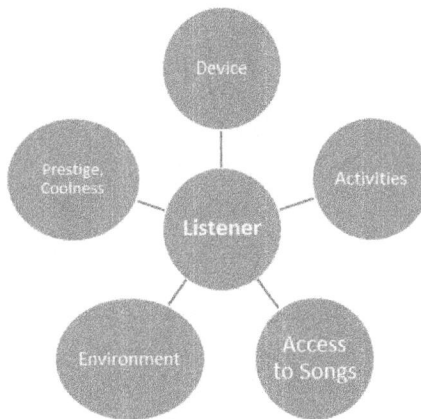

**Figure 16.** Apple Structure [39]

The iPod (and in fact most Apple products) was developed using similar innovative Systems Thinking focused on the user's *experience* as opposed to the product. It is natural to ask if other products should be considered through this innovative Systems Thinking "experiencing" versus "owning" lens:

- The automobile as a stand-alone product versus the *car buying and owning experience*. Car ownership involves car purchase, registration, annual inspections, maintenance, and disposal or trade-in as well as insurance. Cars wear out and new technology renders older models obsolete. There is no good reason that car dealers could not provide all

203

these services for a fixed monthly fee. Some dealerships have already started down this path with service areas that provide free meals, entertainment, and drop-off services. Several manufacturers (Volvo, Cadillac, BMW) have adopted new "subscription services" in which a fixed monthly fee is paid by the user to cover lease, insurance, maintenance, and other expenses [46].

- Coffee. Is it the coffee itself, or is it the *coffee-drinking experience?* Starbucks (and others) attracts clients to not only buy and drink coffee, but to enjoy the coffeehouse experience, with free wi-fi, comfortable seating, and even fireplaces in some establishments.

- Clothes versus the clothes *buying and owning experience.* Clothes use involves clothing selection, travel to a store, fitting, matching, laundering and pressing, repair, and disposal. While many do not mind (or even enjoy) these activities, some do not. Enterprising Systems Thinking businesses could assume all these functions for a fixed monthly fee, thus providing a clothes use *experience* in which the clothes themselves are merely adjuncts.

- Flat Panel TV versus the *home entertainment system experience.* To many people, the home theater component selection, purchase, matching, interconnection, and set up is a harrowing experience. A Systems Thinking approach by subscription TV service providers (Comcast, Verizon, DirecTV, DishTV, etc.) would dictate that these onerous functions be included in the monthly subscription. This would ensure that users would always have the latest equipment set up and functioning optimally to receive the provider's streaming content. It would be similar to a razor-blade or inkjet printer business model in which the asset (in this case the TV and associated hardware) is provided free or near-free to encourage the user to consume the razor blades or ink (in this case the streaming services).

- Home Ownership. Certainly many houses are rented today; however house *buyers* must assume the responsibility for lawn and yard maintenance, utilities, snow removal, pool maintenance, insurance, and all the other onerous responsibilities that come with home ownership. A Systems Thinking approach to the home owning experience would bundle these items in with monthly mortgage payments such that the owner pays one monthly fee for *all* home owning tasks and services, which are then provided by the mortgage company or their representatives.

There are other examples for which close inspection (from a Systems Thinking perspective) reveals that the *product* is really just a part of a *User Experience System.*

## 3. Procedure

The Systems Engineering procedure is described well by Blanchard and Fabrycki [47] and by Kossiakoff et al. [2]. The principle steps are:

1. Regional Architecture/Scoping the Problem
2. Feasibility Study/Concept Exploration
3. Concept of Operations
4. System Requirements
5. High-Level design
6. Detailed Design
7. Software/Hardware Development/Field Installation
8. Unit Device Testing
9. Subsystem Validation
10. System Verification and Deployment
11. System Validation
12. Operations and Maintenance
13. Changes and Upgrades
14. Retirement/Replacement

Many of the engineering and design failures described earlier in this paper were the result of failures to bound the system properly, specifically interactions with environmental factors, and failures to adequately address planned and unplanned interactions among system components themselves and between system components and the environment. Those failures were the result of inadequate scoping of the problem (Step 1 above), which is directly related to the Systems Thinking concepts of Holistic Perspective/Proper Definition of System Boundaries and a Focus on Relationships. If the boundaries and interactions of a system with its environment are not defined properly at the onset of the Systems Engineering process, each of the subsequent steps in that process will be focused on incomplete or inaccurate requirements. The high-level and detailed design of the system will not take into account critical interactions of the system with its environment. Moreover, the Verification and Validation of the system will be incomplete and not consider those interactions of the system with its environment, which could result in catastrophic failure of the system.

To avoid similar failures from arising in the future, a holistic perspective must be taken at the beginning of the Systems Engineering lifecycle, and revisited often throughout that lifecycle. A wide range of relationships among many of the system components themselves and the environment in which the system is intended to operate must be considered. That range of relationships should also include organizational, political, economic, environmental, ethical, and sociological factors. The tools described above are instrumental in achieving this.

Finally, many engineered products (automobiles, appliances, tools, entertainment devices, clothing, prepared foods, engineered homes, etc.) are merely components of a *User Experience System*. Systems Thinking requires that engineers pay attention to all aspects of this system when designing products.

## 4. Conclusions and Recommendations

Engineers can sometimes be excused for problems that have never before been observed. But we should learn from our mistakes. The problems associated with several of the examples listed in this paper had been previously observed and documented, but that information was either not researched, ignored, or (more likely) available in an organized, actionable way. Systems Thinking helps organize some historical engineering and design errors into a set of principles that can be beneficial and actionable. Taking a holistic view and focusing on relationships (as opposed to system components) can minimize the chances of engineering and design gaffs. In this paper, we analyzed 12 Systems Engineering failures and found that they could be categorized into 4 types of Systems Thinking errors: failure to identify relevant environmental factors, failure to understand that most complex problems cannot be solved by purely technological means, failure to adequately address both planned and unplanned interactions among the system components themselves and between system components and the environment, and failure to recognize that many products are actually components of a *User Experience System*. Several tools are available to address these issues:

1.  Early in the design process, system engineers must ensure that they have captured all relevant components of the system, as well as the suprasystem in which the system of interest resides. Environmental factors such as wind, insolation, temperature, and the potential for dramatic incidents such as earthquakes, tsunamis, and hurricanes are notorious for being overlooked. The System Breakdown Structure (SBS) is an excellent tool for this.
2.  Once the relevant components of the system of interest and its suprasystem have been properly identified, all relevant relationships must be identified. The System Interrelationship Matrix (SIS) is an excellent tool for this. Stock-and-Flow diagrams and Causal Loop diagrams may also be helpful.
3.  Having identified all relevant system components and interrelationships, systems engineers must then realize that very few complex problems have purely technological solutions. Engineers also must consider the sociological, psychological,

ethical, political, cultural, and economic factors that may impact the success of a complex system.

4. Finally, in designing complex systems, engineers must understand that "products" are very often not stand-alone devices or systems, but instead part of a *user experience system* that comprises the user, the environment, aesthetics, psychological factors, system acquisition, maintenance, upgrades, and disposal. Failure to address these ancillary factors may cause the system to fail, either technically or commercially.

We recommend that these tools be applied during the first few steps of the standard System Engineering design process to avoid catastrophic Systems Engineering failures in the future.

## References

1. Monat, J. P., and Gannon, T.F., "What Is Systems Thinking? A Review of Selected Literature Plus Recommendations," *Am. J. of Systems Science*, 4:2, 2015

2. Kossiakoff, Alexander, Sweet, William, Seymour, Samuel, and Biemer, Steven, *Systems Engineering: Principles and Practice*, 2nd Ed., Wiley, Hoboken, 2011

3. Lawson, Harold, *A Journey Through the Systems Landscape*, College Publications, London, 2010, 6

4. Boyd, J. R., "An Organic Design for Command and Control, A Discourse on Winning and Losing," Unpublished lecture notes. Maxwell Air Force Base, Alabama Air University, 1987

5. Shewhart, W. A., *Statistical Method from the Viewpoint of Quality Control*. U. S. Department of Agriculture. Dover, 1939, 45

6. Kasser, J. and Mackley, T. (2008), Applying systems thinking and aligning it to systems engineering. INCOSE International Symposium, 18: 1389–1405. doi:10.1002/j.2334-5837.2008.tb00886.x

7. Godfrey, Patrick, "What is Systems Thinking? How does it relate to Systems Engineering?" INCOSE UK zGuide 7, Issue 1, March 2010, https://incoseonline.org.uk/Program_Files/Publications/zGuides_7. aspx?CatID=publications&SubCat=zGuides, accessed 10 April 2018

8. Pyster, A., D. Olwell, N. Hutchison, S. Enck, J. Anthony, D. Henry, and A. Squires (eds). 2012. *Guide to the Systems Engineering Body of Knowledge (SEBoK) version 1.0*. Hoboken, NJ: The Trustees of the Stevens Institute of Technology ©2012. Available at: http://www.sebokwiki.org., 98-125

9. INCOSE, Systems Engineering Handbook, 4th Edition, INCOSE-TP-2003-002-04-2015, 2015, 18-21

10. Burge, Stuart E., "Systems Engineering: Using Systems Thinking to Design Better Aerospace Systems," *Encyclopedia of Aerospace Engineering*, John Wiley & Sons, Ltd., 2010

11. Frank, Moti, "Engineering Systems Thinking: Cognitive Competencies of Successful Systems Engineers", Procedia Computer Science Volume 8, 2012, Pages 273-278

12. Billah, K. Yusuf, and Scanlan, Robert H., "Resonance, Tacoma Narrows bridge failure, and undergraduate physics textbooks," *American Journal of Physics* **59** (2), 1991, 118-124

13. Petroski, Henry, *Design Paradigms---Case Histories of Error and Judgment in Engineering*, Cambridge University Press, New York, 1994, 156-165

14. Gaal, Rachel, "This Month in Physics History, November 7, 1940: Collapse of the Tacoma Narrows Bridge," APS News 25 (10), November 2016, https://www.aps.org/publications/apsnews/201611/physicshistory.cfm, accessed 14 August 2018

15. Dallard, P. et al. "The London Millennium Footbridge," *Structural Engineer*, **79**:22, 20 November 2001. pp. 17–35

16. Josephson, Brian, "Out of Step on the Bridge," Guardian Letters, 14 Jun 2000, https://www.theguardian.com/theguardian/2000/jun/14/guardianletters3, retrieved 7 May 2018

17. Strogatz, Steven, Daniel M. Abrams, Allan McRobie, Bruno Eckhardt, & Edward Ott, "Theoretical Mechanics: Crowd Synchrony on the Millennium Bridge," *Nature*, Vol. **438**, 2005, 43-44

18. Bachmann Hugo, and Ammann, Walter, "Vibrations in Structures Induced by Man and Machines", *Structural Engineering Document 3e*, International Association for Bridge and Structural Engineering (IABSE), Ch. 2: Man-Induced Vibrations, and Appendix A: Case Reports, 1987

19. Fujino, Yozo, M Pacheco, Benito M., Nakamura, Shun-ichi, and Warnitchai, Pennung, "Synchronization of Human Walking Observed during Lateral Vibration of a Congested Pedestrian Bridge", *Earthquake Engineering and Structural Dynamics*, **22**, 1993, 741-758 1993

20. Wernimont, E., M. Ventura, G. Garboden and P. Mullens, "Past and Present Uses of Rocket Grade Hydrogen Peroxide," General Kinetics, LLC Aliso Viejo, CA 92656, 2018

21. Duddu, Praveen, "The World's Deadliest Torpedoes," Naval Technology, 8 June 2014, https://www.naval-technology.com/features/featurethe-worlds-deadliest-torpedoes-4286162/ accessed 7 May 2018

22. Guardian, "What Really Happened to Russia's 'Unsinkable' Sub", 4 August 2001, accessed 7 May 2018

23. Faulconbridge, Guy (3 December 2004). "Nightmare at Sea". Moscow Times. Archived from the original on 28 February 2014. Retrieved 22 February 2014

24. Garfield, Leanna, "The 'Death Ray Hotel' Burning Las Vegas Visitors Came Up with a Simple Fix," Business Insider, 30 June 2016, http://www.businessinsider.com/the-vdara-death-ray-hotel-is-still-burning-people-in-las-vegas-2016-6, accessed 7 May 2018

25. Maimon, Alan, "Vdara Visitor: 'Death Ray' Scorched Hair," *Las Vegas Review-Journal*, 24 September 2010, https://www.reviewjournal.com/news/vdara-visitor-death-ray-scorched-hair/, accessed 7 May 2018

26. Hodge, Damon, "Reflective "Death Ray" Torments Vegas Sunbathers," Reuters "Environment," 1 October 2010, https://www.reuters.com/article/us-lasvegas-deathray-idUSTRE6904AR20101001, Accessed 7 May 2018

27. Smith-Spark, Laura, "Reflected Light from London Skyscraper Melts Car," *CNN*, 3 September 2013, http://edition.cnn.com/2013/09/03/world/europe/uk-london-building-melts-car/index.html, accessed 7 May 2013

28. BBC, "Who, What, Why: How Does a Skyscraper Melt a Car?" 3 September 2013, http://www.bbc.com/news/magazine-23944679, Accessed 7 May 2018

29. YouTube, "Man Fries Egg in London Skyscraper Heat," TheNetFrancesco, 3 September 2013, https://www.youtube.com/watch?v=TGnZz8-PvdY, accessed 7 May 2018

30. Sherwin, Adam, "Walkie Talkie City Skyscraper Renamed Walkie Scorchie after Beam of Light Melts Jaguar Car Parked Beneath It," The Independent, https://www.independent.co.uk/arts-entertainment/architecture/walkie-talkie-city-skyscraper-renamed-walkie-scorchie-after-beam-of-light-melts-jaguar-car-parked-8794970.html, 3 September 2013, accessed 7 May 2018

31. Washington Post News Service, "London's 'Fryscraper' Draws Crowd on Hottest Day, *Mississauga News*, 6 September 2013, https://www.mississauga.com/news-story/4067822-london-s-fryscaper-draws-crowd-on-hottest-day/, accessed 7 May 2018

32. Rhee, Joseph, Ross, Brian, Ferran, Lee, and Hosford, Matt, "Toyota Recalls 2.17 Million Vehicles Over Pedal Entrapment," ABC News, 24 February 2011, https://abcnews.go.com/Blotter/toyota-recalls-21-million-vehicles-floor-mats-pedals/story?id=12989254, accessed 7 May 2018

33. Evans, Scott, and MacKenzie, Angus, "The Toyota Recall Crisis: A Chronology of How the World's Largest and Most Profitable Automaker Drove into a PR Disaster," Motor Trend, 27 January 2010, http://www.motortrend.com/news/toyota-recall-crisis/, accessed 7 May 2018

34. Cramer, Matt, "11 Terrible Automotive Engineering Decisions," https://jalopnik.com/5570865/11-terrible-automotive-engineering-decisions/, 2010, accessed 17 August 2018

35. Feldt, Allan G., 1986, Chapter 2: General Systems Theory, http://www.holisticwisdom.org/hwpages/chapt%202%20-%20GST.htm, accessed 18 June 2018

36. Jon, "R.I.P Microsoft Zune (2006-2011)," Methodshop, https://www.methodshop.com/2011/03/microsoft-zune-rip.shtml, 2011, accessed 13 August 2018

37. Norman, Don, "Designing for People--Systems Thinking: A Product is More than the Product," *Interactions* **16** (5), Sept-Oct 2009

38. Johns Hopkins Whiting School of Engineering, "System Context Diagrams," 1 February 2014, https://ep.jhu.edu/about-us/news-and-media/systems-context-diagrams

39. Monat, J. P., and Gannon, T.F., *Using Systems Thinking to Solve Real-World Problems*, College Publications, London, 2017

40. The Systems Thinker, "Guidelines for Drawing Causal Loop Diagrams," 22 (1), Pegasus Communications, February 2011, https://thesystemsthinker. com/wp-content/uploads/pdfs/220109pk.pdf

41. Kim, Daniel H., *Introduction to Systems Thinking*, Pegasus Communications, Waltham, MA, 1999, 5-10

42. Federal Aviation Administration, System Engineering Functional N2 Diagram, 2006

43. Bechtel Jacobs Company, "UF6 Cylinder Project Systems Engineering Management Plan," U. S. Dept. of Energy, July, 1998, http://www. oakridge.doe.gov/duf6disposition/KTSO-017.pdf

44. U.S. Department of Defense, Chief Information Officer, "DoDAF Viewpoints and Models, Systems Viewpoint, SV-3: Systems-Systems Matrix," https://dodcio.defense.gov/Library/DoD-Architecture-Frame-work/dodaf20_sv3/, accessed 11 June 2018

45. Wagenhals, Lee, and Levis, Alexander, "C4ISR Architectures and their Implementation Challenges, Lecture 4: A C4ISR Architecture Framework Implementation Process Case Study," INCOSE, http://slideplayer. com/slide/701808/, accessed 11 June 2018

46. Hyatt, Kyle, "A Guide to Car Subscriptions, a New Alternative to Buying and Leasing," Road/Show by CNET, 14 June 2018, https://www.cnet. com/roadshow/news/new-car-subscription-service-guide-buying-leasing-2018/ accessed 27 June 2018

47. Blanchard, Benjamin and Fabrycky, Wolter, *Systems Engineering and Analysis*, 5th Ed., Prentice Hall, Upper Saddle River, NJ, 2011.

# USING SYSTEMS THINKING TO ANALYZE ISIS

Jamie P. Monat and Thomas F. Gannon

Systems Engineering Program
Worcester Polytechnic Institute
Worcester, MA 01609
USA

*Reprinted from American Journal of Systems Science*, Vol. 4 No. 2, 2015, 36-49

## Abstract

Systems Thinking can be used to address complex socio-economic problems, predict behaviors, and understand the seemingly illogical actions of individuals, countries, and organizations such as ISIS. It focuses on relationships among system components as well as on the components themselves. In this paper, we apply Systems Thinking tools (including the Iceberg Model, causal loop diagrams, stock-and-flow diagrams, and dynamic modeling) to analyze ISIS's beliefs, goals, needs, and appeal, and to suggest new strategies for dealing with ISIS. We conclude that a Systems Thinking analysis leads to approaches that are very different from typical linear thinking solutions, such as "bomb them back to the Stone Age," and suggest alternative approaches for dealing with ISIS. These include waging a non-military but a socio-economic war against ISIS using social media, moving away from a policy of forcibly imposing democracies, rethinking the U. S.'s role as the world's policeman, destroying ISIS's sources of revenue, encouraging ISIS's Middle Eastern neighbors to fight the land war (with the U. S. serving only in an advisory capacity), addressing the root causes of ISIS's appeal, and preventing both Iran and ISIS from developing nuclear capabilities at all costs.

## Introduction and Background

There are various current trends in and perspectives on Systems Thinking (Anderson and Johnson, 1997; Boardman and Sauser, 2008; Kauffman, 1980; Kim, 1999; Meadows, 2008; Richmond, 2004; Senge, 1990 and 2006; Weinberg, 2001. For a more extensive literature review and summary, see Monat and Gannon, 2015). From these references we have concluded that *Systems Thinking* is a perspective, a language, and a set of tools that can be used to address complex socio-economic issues. Specifically, Systems Thinking is the opposite of linear thinking. It is a holistic approach to analysis that focuses on the way a system's constituent parts interrelate and how systems work over time and within the context of larger systems. Systems Thinking recognizes that repeated events or patterns are derived from systemic structures which, in turn, are derived from mental models. It recognizes that behaviors derive from structure, it focuses on relationships rather than components, and it recognizes the principles of self-organization and emergence. Specific Systems Thinking tools include systemigrams, system archetypes, main chain infrastructures, causal loops with feedback and delays, stock and flow diagrams, behavior-over-time graphs, computer modeling of system dynamics, Interpretive Structural Modeling (ISM), and systemic root cause analysis (Monat and Gannon, 2015).

ISIS, the Islamic State of Iraq and Syria, is a jihadist organization of Sunni Muslims who have been gaining support and recruits in their effort to achieve world domination, killing and maiming thousands of non-believers in the process. Their behavior has seemed self-destructive and illogical inasmuch as many of their actions have united their Middle Eastern neighbors (and in fact, much of the world) against them. Several experts and political scholars have attempted to analyze ISIS.

Princeton alumnus William McCants directs the Project on U.S. Relations with the Islamic World at the Brookings Institution, teaches at Johns Hopkins University, and was a senior advisor for countering violent extremism at the U.S. Department of State. His new book, *The ISIS Apocalypse: The History, Strategy, and Doomsday Vision of the Islamic State,* provides a behind-the scenes look at ISIS's history (focusing on the years 2003-2015) and motives, gleaned from private e-mails, text messages, and communiques among ISIS leaders, followers, and opponents.

Shadi Hamid is a Senior Fellow in the Center for Middle East Policy at the Brookings Institute and expert on U. S. Relations with the Islamic World. In his article "Is There a method to ISIS's Madness," Hamid explores ISIS's rationality as well as that of the West. He discusses the Caliphate's need to hold and expand territory and speculates on future ISIS behavior as they lose both territory and morale. He also speculates on the reasons that ISIS's enemies have thus far failed to unite and on ISIS's possible objectives.

Loren Thompson's article "Five Reasons the ISIS Fight Isn't About Islam" draws clear distinctions between ISIS and Islam. In it, he explains how ISIS is fundamentally different from other jihadi groups and discusses the appeal of ISIS to young, disenfranchised men. Dr. Thompson is Chief Operating Officer of the Lexington Institute and was Deputy Director of the Security Studies Program at Georgetown University; he has taught both at Georgetown and at Harvard's Kennedy School of Government.

David Von Drehle, correspondent for *Time* magazine, wrote "The War on ISIS" in March of 2015. For this article, Von Drehle interviewed ISIS sympathizer Anjem Choujary, Deputy National Security Advisor Ben Rhodes, Aron Lund (editor of "Syria in Crisis" for the Carnegie Endowment for International Peace), and Fawaz Gerges, Emirates Chair in Contemporary Middle East Studies at London's School of Economics. The article describes some of the pitfalls associated with imposing democracy, as well as how the U. S. may have contributed to the rise of ISIS.

One of the best analyses was conducted by Graeme Wood and published in *The Atlantic* in March of 2015. For his article, Wood interviewed Princeton scholar Bernard Haykel, a leading expert on ISIS; Peter R. Neumann, Director of the International Centre for the Study of Radicalisation and Political Violence and professor of Security Studies at King's College, London; Australian Musa Cerantonio, widely regarded as one of the most important ISIS spiritual authorities; Londoners Anjem Choudary, Abu Baraa, and Abdul Muhid, all former members of the banned Islamist group "Al Muhajiroun," and Salafi imam Breton Pocius from Philadelphia. Based on these interviews, Wood was able to develop a fairly clear picture of ISIS's motives and objectives.

In the sections of this paper that follow, we have attempted to synthesize the views of these political experts and use them together with Systems Thinking to understand ISIS's seemingly illogical behavior and, in fact, to address the issue. To analyze this situation, it is helpful to summarize ISIS's beliefs, goals, needs, and appeal, as articulated by these political experts.

**ISIS Beliefs.** ISIS believes that the Koran is a manual for war and that it must be followed literally. Slavery, crucifixion, beheadings, stonings, amputation, and rape of infidel women are all mandated by the Koran. ISIS further believes that infidels must be treated without mercy (an infidel is anyone who does not follow the Koran literally). But there are different kinds of infidels: pagans (such as Christians and Jews) must be enslaved; but the worst kind of infidels are apostate Muslims, who must be put to death. Included in this category are Al-Qaeda and all Shia. ISIS also believes that democracy, elections, national borders, alcohol, drugs, and enforcing any law not made by God are immoral and excommunicable; and that it is apostate to recognize any authority other than God. Interestingly, ISIS's principal enemies are apostate Muslims, not Christians, Jews, or the U.S.; but ISIS does believe that the U.S. wants to eradicate Muslims.

ISIS declared a Caliphate on June 29th, 2014, designating Abu Bakr al-Baghdadi as its Caliph. This is the first Caliphate declared in the Muslim world in over 1,000 years. ISIS believes that the Caliphate is a vehicle for salvation. To go to paradise, a Muslim must pledge allegiance to the Caliph. There will be only 12 legitimate Caliphs (Baghdadi is the 8th.)

ISIS believes that the path to paradise is via Jihad (Holy War), which must be waged at least once per year. Final Sunni victory will occur at a battle in Jerusalem at which the anti-Messiah (from Iran) will kill all except 5,000 of the faithful, who will be rescued at the last minute by (interestingly) Jesus. Prior to this final showdown, there will be a period of Islamic conquest led by the Mahdi, a messianic figure destined to lead the Muslims to victory before the end of the world. The period of Islamic conquest will be initiated by a victory at Dabiq, in northern Syria, where the armies of Rome will meet the armies of Islam. ISIS has already stationed people at Dabiq and want to wage the Battle of Dabiq as soon as possible.

**ISIS Goals.** The ultimate goal of ISIS is world domination leading to the end of days, at which time the most faithful will ascend to paradise. But there are several sub-goals required to achieve this endgame. The first is a Caliphate with Abu Bakr al-Baghdadi as Caliph (a "Caliphate" is an Islamic state; a form of Islamic government led by a Caliph who is a political and religious leader and a successor (Caliph) to the Islamic prophet Mohammed. His power and authority are absolute.) Baghdadi is the first declared Caliph in more than 1,000 years. Before the apocalypse, ISIS wants to return civilization to a 7th century legal environment with no borders, after forcible expansion into non-Muslim countries, and to purify the world by killing vast numbers of people who are apostate, per Takfiri doctrine. To further this goal, ISIS does not enter into long-lasting (meaning >10 years) peace treaties. Another ISIS goal is free housing, food, clothing, and health care for all, which may explain its appeal to Sunni Muslims who do not have these now.

It is important to note that as ISIS's territory and populace expand, its need for resources to support this goal will expand dramatically. Finally, to hasten the end of times, ISIS wants to draw the U. S. into a war and accelerate the battle of Dabiq, which represents the start of the countdown to the apocalypse. ISIS's outrageous acts of beheadings, immolations, and other atrocities are designed to terrorize its enemies into either capitulation or outright war and thus serve this goal.

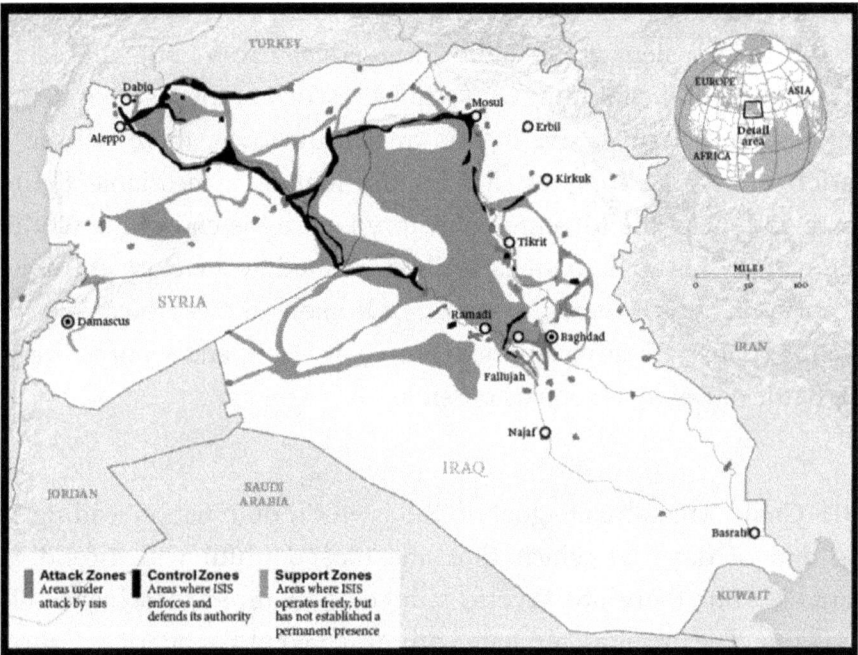

**Figure 1. ISIS Territories as of December 2014** (Map developed by the Institute for the Study of War and published in *The Atlantic*, March 2015)

**ISIS Needs.** To achieve its goals, ISIS needs a Caliph (and 4 more after Baghdadi), territory it controls, and a continual *expansion* of territory it controls. This need to take and hold territory is a vulnerability. (From the map in Figure 1, it seems clear that ISIS is pushing outward in a ring centered west of Kirkuk. With only 30,000 soldiers, as this perimeter grows, the resources required to retain and control the center of the ring will stretch thin. This represents a vulnerability and an opportunity for ISIS's enemies to infiltrate and fight ISIS from within its own perimeter.) ISIS also needs money and resources for food, clothing, healthcare, weapons, and ammunition in an ever-growing geography and populace. It also needs continuous victories and growth to maintain internal support and an influx of new recruits (another vulnerability). ISIS functions on a strong command/control management structure in which the Caliph holds ultimate power and authority. Finally, ISIS needs victory at Dabiq and Jesus Christ, who is prophesied to be the savior of the Islam faithful at Jerusalem at the time of the apocalypse.

**ISIS's Appeal:** Why do people join ISIS? Most normal people find it barbaric, inhumane, and vile. Yet it holds great appeal to some. To some truly pious Sunni Muslims, ISIS is appealing because of its strict adherence to Sharia and because of its promise of paradise to the faithful. To adventure-seekers, people who seek a cause to fight for, and those who seek power and control, ISIS represents an answer. To Muslims and others world-wide who resent the U.S.'s self-appointed role as policeman of the world, ISIS represents an opportunity to correct that. However, most significantly, one needs to understand Maslow's hierarchy. To the 30,000 soldiers that fight for ISIS, the $400 monthly salary may be significantly better than what they could earn doing anything else. To the disenfranchised, to those whose lives are miserable, and to the growing number of unemployed youth in Arab countries who perceive their governments as corrupt, unaccountable, and coercive (and controllers of jobs and the economy), ISIS appears to be more fair, just, and moral, and to offer more hope for a good life. These people aren't joining/supporting ISIS from some high-level philosophical ideological perspective. They are joining because under ISIS, for the first time, they can eat, stay warm, and see a doctor.

## Systems Thinking Analysis

**A Systems Thinking Approach to Warfare:** In *Blink*, Malcolm Gladwell says, "Conflict in the future would be diffuse. It would take place in cities as often as on battlefields, be fueled by ideas as much as by weapons, and engage cultures and economies as much as armies." As one US Joint Forces Command (JFCOM) analyst puts it: "The next war is not going to be military on military. The deciding factor is not going to be how many tanks you kill, how many ships you sink, and how many planes you shoot down. The decisive factor is how you take apart your adversary's *system*. Instead of going after war-fighting capability, we have to go after war-making capability. The military is connected to the economic system, which is connected to their cultural system, to their personal relationships. We have to understand the links between all those systems."

With respect to ISIS, we should not be fighting a military war, but a socio-economic one.

ISIS needs huge sums of money to fund its military and social programs. Its principal sources of income are oil (which is smuggled out of the country), wheat, taxes and extortion from its own populace, and ransom for hostages. Estimates are that ISIS is worth $2 billion and generates about $30 million/month from oil, $30 million/month from extortion/taxes, and $2 million/month from hostage ransoms (Bajekal, Bronstein). ISIS's cultural appeal (based upon both fairness according to Allah and guaranteed food, clothing, and medical care to yield a better life) renders it susceptible to criticism and disillusionment if those promises are not fulfilled, and if the populace grows to resent the extortion and increased tax rates. As oil production infrastructure has been targeted, ISIS has relied more upon taxes and extortion to fund its coffers.

## Iceberg Models

The Iceberg Model (Kim, 1999; Meadows, 2008; Anderson and Johnson, 1997; Senge et al., 1994) is an excellent systems thinking tool. It is useful for understanding how mental models give rise to systemic structures which, in turn, yield patterns of events. The events and patterns are visible to us, but the mental models and structures are often hidden and must be discovered.

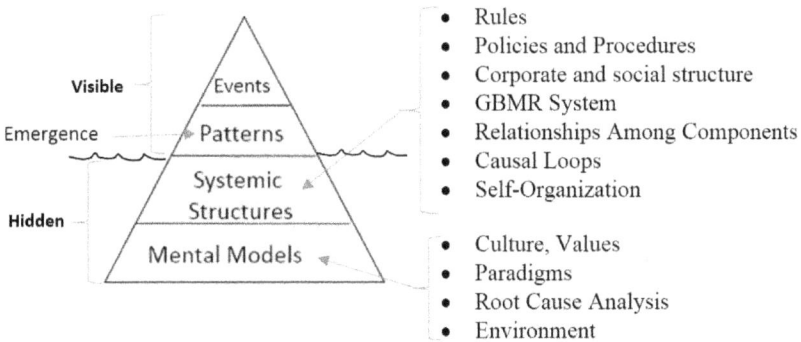

**Figure 2. The Iceberg Model.** "GBMR System" means the organization's Goals-Behaviors-Metrics-Rewards system which is used to incent organization members to achieve specific objectives via suitable performance measurements and rewards.

Table 1 below lists some relevant mental models and the structures, patterns, and events that result based on our analysis of ISIS's beliefs, goals, needs, and appeal. Each row in the table represents a different Iceberg Model.

## Table 1. Iceberg Models Relevant to ISIS.

| Iceberg Model Number | Mental Model | Resulting Structure | Pattern | Events |
|---|---|---|---|---|
| 1 | It is preferable to die a martyr (and ascend to paradise) than to continue the life I have been living. | ISIS, infrastructure for war, incentive for suicide bombings | War | Battles, suicide bombings |
| 2 | I will be 1 of the 5,000 ultimate survivors, if I am 1 of the 5,000 who follow the Koran most closely. | Religious and political structure of ISIS | Repeated Murders, rapes, be-headings, amputations, annual Jihad in the name of the Koran | Murders, rapes, beheadings, amputations, annual Jihad |
| 3 | If there will be 12 Caliphs before the end of time and we are on #8, then we had better get going! | The Cali-phate, the propaganda, political, and social media efforts to engage the US and the rest of the world in war | Repeated outrageous provocations to expedite the Battle of Dabiq and the end of times | Live immo-lations, be-headings of innocents, enslavement, other sen-sational, provocative acts |
| 4 | The U.S. is a big, powerful bully that wants to eradicate Islam. | Political, Military, and Social structures that denigrate the U. S. | Increased opposition to U.S. inter-vention in foreign af-fairs, repeat-ed anti-U.S. demonstra-tions and activities, | Anti-U.S. dem-onstrations and activities, attacks on U.S. embassies, military bases, ships and convoys |

| | | | attacks on U.S. embassies and military assets abroad | |
|---|---|---|---|---|
| 5 | ISIS economic and government policies are more just, fairer, and more moral than my current government's. | Religious and political structure of ISIS; incentive structure for ISIS recruits | Recurring cycles of oppression and revolts between competing religious and political factions | Recruitment of ISIS fighters and civilians |
| 6 | Democracy is the best government and the US has the moral obligation to impose it. | The U. S. military and political machine | Repeated intervention in foreign affairs and forced installation of democratic governments | Intervention in countries we don't like and forced installation of democracies. |
| 7 | The US has a moral obligation to be the world's policeman. | The U. S. military and political machine | Repeated intervention in foreign affairs | Intervention in foreign affairs |
| 8 | Before I joined ISIS, my life meant nothing. Now, even if I die young, my life will have mattered. | ISIS, infrastructure for war, incentive for suicide bombings | War | Battles, suicide bombings |
| 9 | The sooner we destroy this world, the sooner we will be welcomed in paradise | Incentive to destroy as much of the world as possible, as soon as possible. | Repeated attempts to acquire nuclear weapons | Nuclear holocaust |

## ISIS vs Iran: 2 Warring Caliphates

ISIS (Sunni) and Iran (Shiite) may have similar goals. One significant difference is that ISIS has already declared a Caliphate while Iran has not. However, Iran's refusal to commit to long-term treaties; steadily increasing influence over western Afghanistan, Iraq, Syria, Lebanon, and Yemen; and recent calls for a "global Caliphate" suggest similar thinking. Indeed, according to Shillman Fellow in Journalism Daniel Greenfield, "Iran views its struggle with the US as a gateway to the apocalypse, the Islamic end of days and the mass murder of non-Muslims." The greatest threat to the world would be a Caliphate with nuclear capability, which would transpire if Iran wins the war against ISIS and develops a nuclear bomb. This implies that the defeat of ISIS is an immense *threat* and that it is to the world's advantage to let ISIS continue to keep Iran in check. It also suggests that Iran must be kept from developing nuclear capabilities at all costs. (ISIS may also be attempting to acquire weapons-grade uranium.)

## Causal Loop Diagrams

Causal loop diagrams (CLDs) show cause-and-effect relationships among system components (Anderson and Johnson, 1997; Goodman et al.; Kauffman, 1980; Richmond, 2004). They are especially useful for understanding the role of feedback, which is very often present (and under-appreciated) in systems. There is a famous causal loop diagram that was developed by the PA Consulting Group describing the war in Afghanistan, shown below in Figure 3.

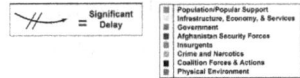

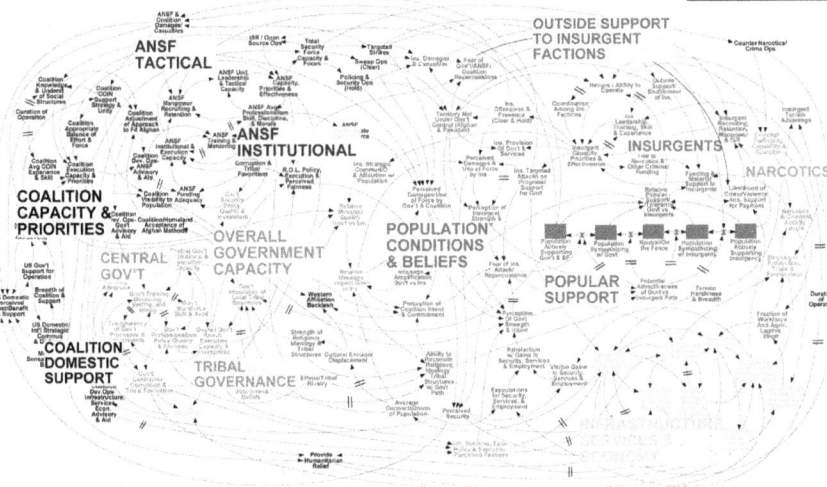

Figure 3. CLD Model of the War in Afghanistan.

Other than demonstrating how complex the situation is there, such busy, complicated diagrams are not particularly useful for understanding cause, effect, and feedback. We believe it is more instructive to develop CLDs for smaller concepts from which some insight may be gleaned. Here we present several CLDs representing background and also future projections for ISIS.

*Background, Historic Loops*

1. The U. S. as World Policeman: A Reinforcing Feedback Loop: A reinforcing (or positive) feedback loop occurs whenever an action or change builds upon itself at an ever-increasing rate. Common examples include interest-bearing savings accounts which continue to add a fixed % interest to a growing principal, and population growth in which constant birth rates applied to steadily-growing populations result in exponential growth. In the current

case, continued intervention by the U. S. in local conflicts will reinforce the resentment of other countries toward the U. S., which will (in turn) yield increased anti-American activity, which will result in more American intervention.

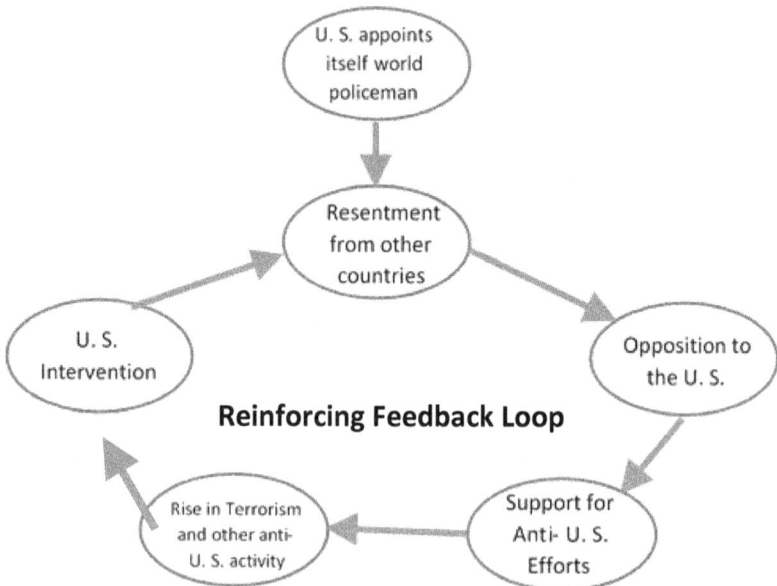

**Figure 4. The U.S. as World Policeman.**

2.  The U. S. as World Policeman: The Addiction Archetype: The addiction archetype is a common negative structure that occurs when an intervener facilitates destructive behavior by an agent, resulting in dependency on the intervener and a loss of accountability by the agent. Common examples are a parent's intervention in a child's homework instead of letting the child err and learn for herself, a poorly-run company depending upon government bailouts instead of getting its own house in order, and a drug or alcohol addict depending upon drugs/alcohol for happiness instead of facing reality. In this case, it is countries depending upon the U. S. to fight their battles, which ultimately de-values freedom and democracy because those countries don't need to fight battles and secure their freedom on their own. Freedom is

devalued and underappreciated when it is just handed to a populace.

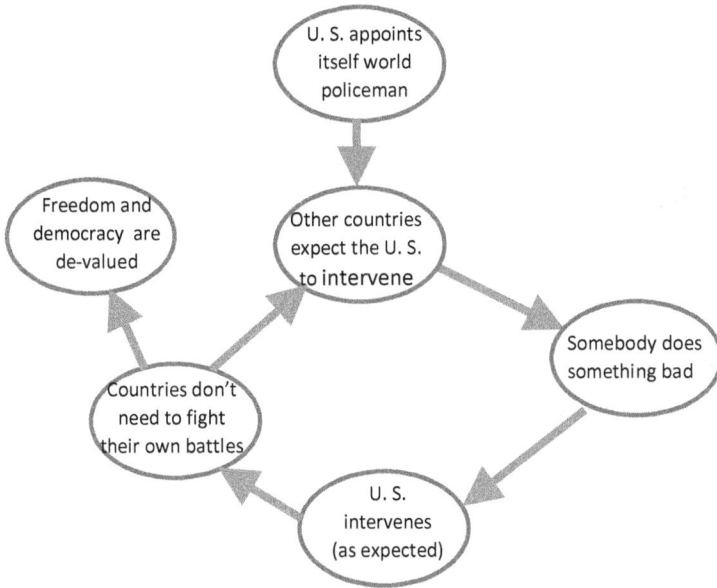

**Figure 5. The Addiction Archetype.**

It is clear from the above CLDs that these reinforcing feedback loops must be broken by discontinuing the use of the "U. S. as World Policemen" mental model and encouraging ISIS's Middle Eastern neighbors to coordinate their efforts to attack and defeat ISIS on their own.

3. <u>The Recurring Cycle of Arab Oppression:</u> Another example of a relevant reinforcing feedback loop is the recurring cycle of oppression among competing political regimes. As one regime is overthrown, another oppressive regime emerges, and the vicious cycle continues.

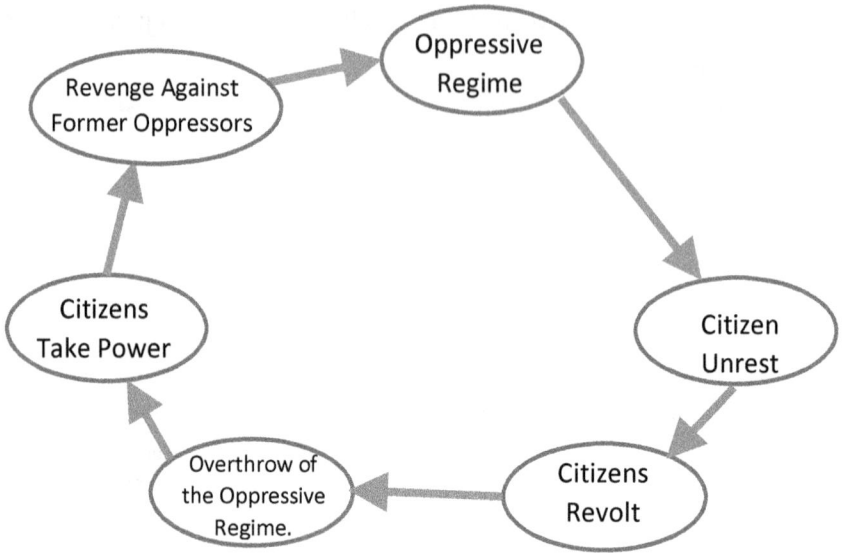

**Figure 6. The Recurring Cycle of Arab Oppression.**

4. <u>How the U. S. Contributed to the Rise of ISIS and Could Repeat this Mistake:</u> As shown in the following diagram, the Addiction Archetype coupled with the recurring cycle of oppression among competing regimes illustrates the rise of ISIS as an unintended consequence of the U.S. intervention in Iraq.

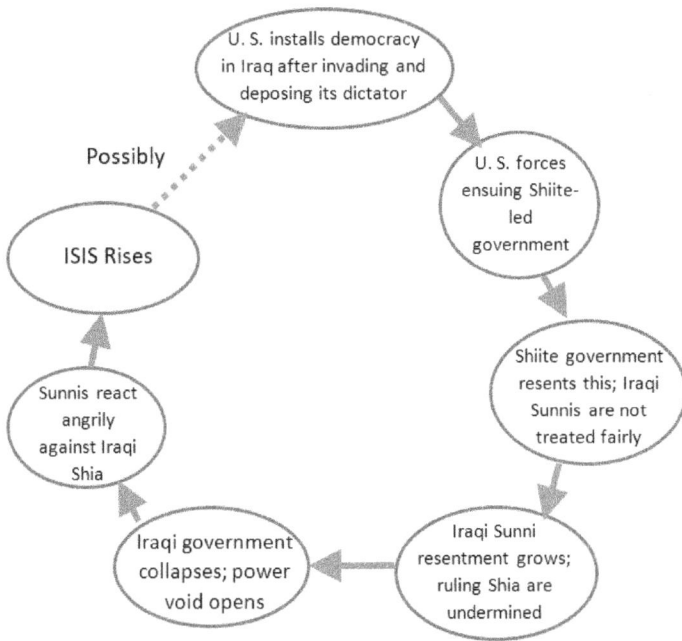

**Figure 7. The Rise of ISIS.**

It is clear from this CLD that this mistake could be repeated in the future if subsequent U. S. intervention is taken to defeat ISIS and impose democracy without understanding the potential unintentional consequences given the region's culture, history, and politics.

*Future Projection Loops*

1. <u>Destruction of ISIS Infrastructure:</u> One approach for accelerating the demise of ISIS is to deny it access to the resources and revenues needed to provide sustenance to its growing populace, which will result in a loss of internal support. As shown below, continued bombing of ISIS oil fields would weaken the ISIS military, reclaim some lost territory, and weaken support for ISIS among its constituents.

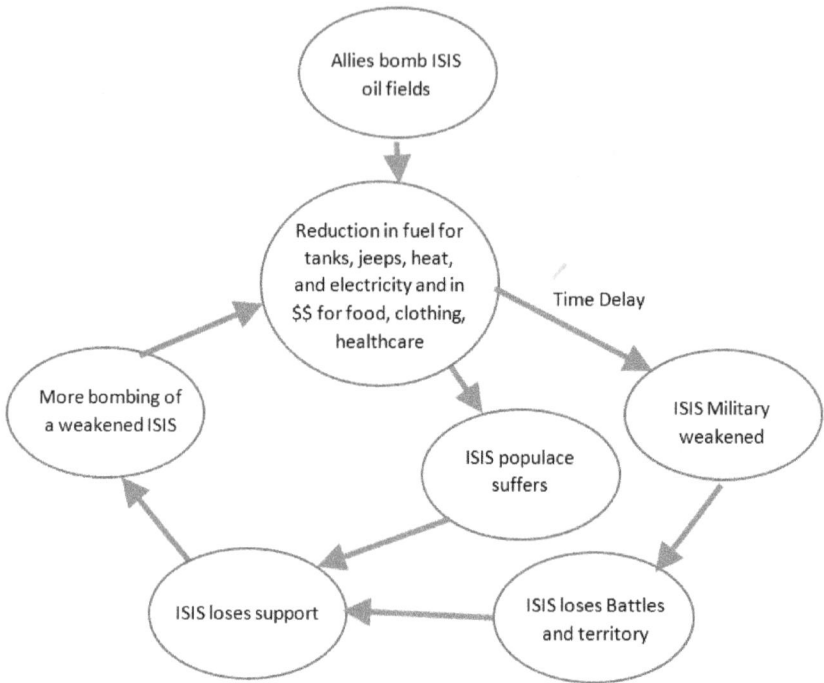

**Figure 8. The Destruction of ISIS Infrastructure.**

2.  <u>Ransom for Hostages:</u> An additional source of revenue for
    ISIS is the payment of ransom in exchange for the release
    of hostages. Success in generating funds from hostage-
    taking encourages the taking of more hostages.

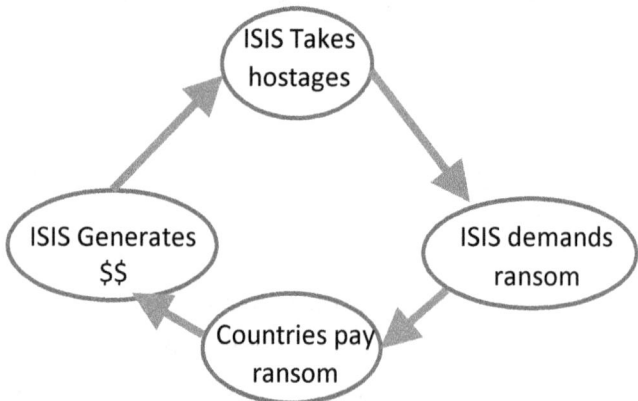

**Figure 9.  ISIS Revenue from Hostage Ransom.**

It is clear from this CLD that the reinforcing feedback loop must be broken by refusing to pay ransom for hostages.

**Stock-and-Flow Diagrams**

A simplified stock-and-flow diagram (Bellinger, 2004; Kim, 1999; Meadows, 2008; Richmond, 2004) may be used to model the amount of resources and revenues ($$) needed as ISIS-controlled territory and population grow. The diagram below was created using iThink dynamic modeling software from isee Systems, Inc., Lebanon, NH (model details are available from the authors). Territory Gain Rate, Revenues ($$) from Oil, and Per Person Costs of Food, Clothing, and Healthcare were set as independent variables in this model.

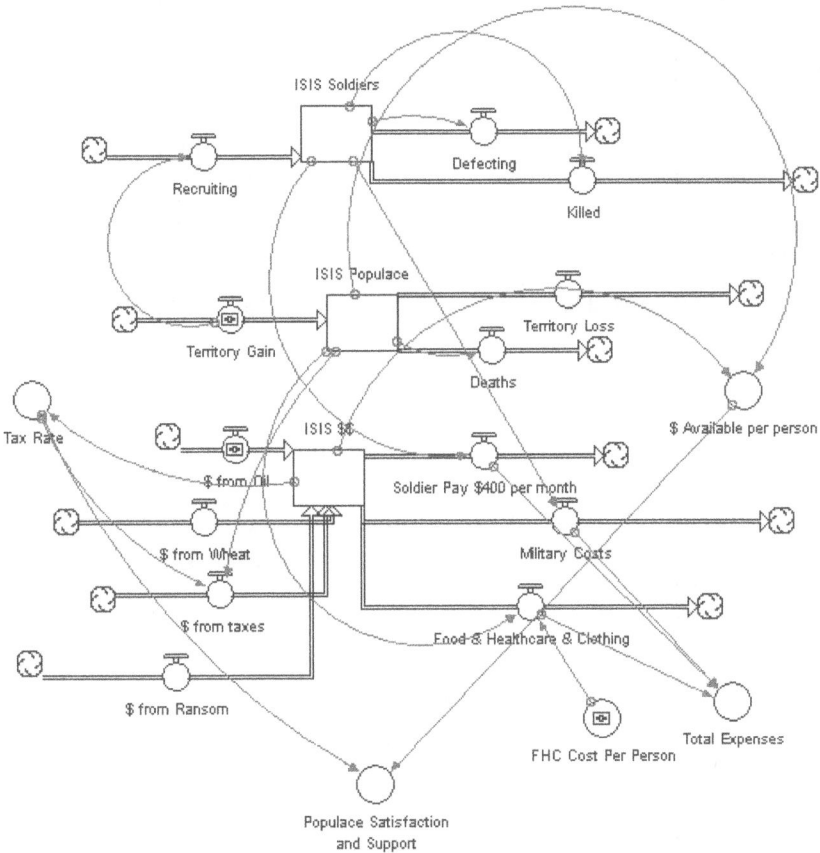

**Figure 10. Dynamic Simulation Model of ISIS Revenue, Resources, Cash Reserves and Popular Support.**

The following three figures are output graphs from the dynamic simulation, and illustrate a dramatic increase in expenses, a decrease in ISIS's cash reserves, and a dramatic decrease in popular support as ISIS gains territory and soldiers.

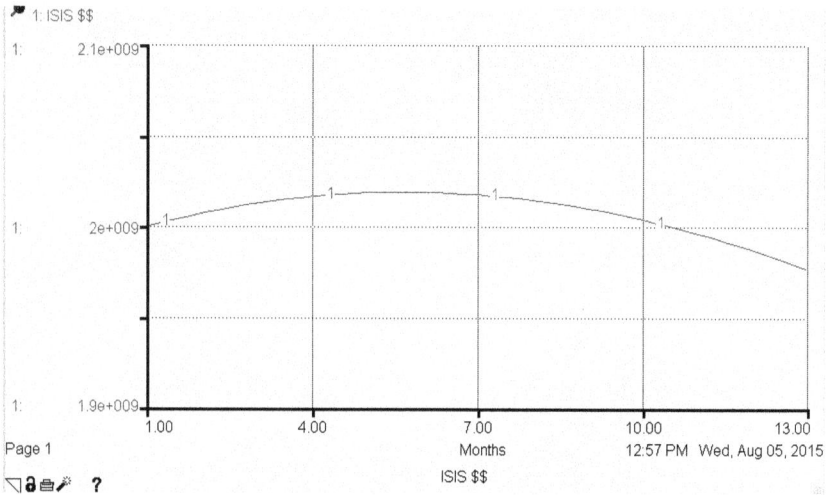

**Figure 11. ISIS Cash Reserves versus Time.**

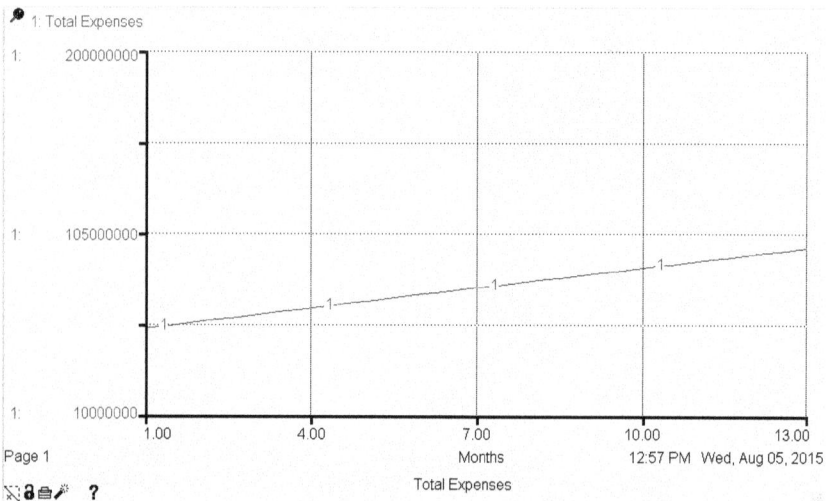

**Figure 12. ISIS Expenses versus Time.**

230

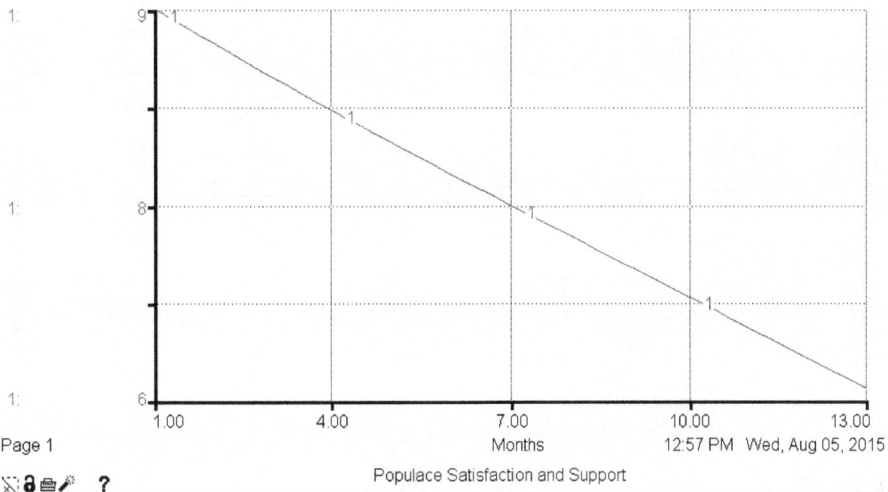

Page 1        Months        12:57 PM  Wed, Aug 05, 2015

Populace Satisfaction and Support

**Figure 13. ISIS Popular Satisfaction and Support versus Time.**

The model reveals that funding; rate of population gain via acquired territory; and food, clothing, and healthcare costs per person are key leverage points of the system. If the ISIS-controlled population increases faster than its revenues, then either ISIS's cash reserves or the services it provides to its people (or both) will decrease, resulting in loss of support. It is clear that destruction of ISIS's oil, wheat, and financial infrastructure while increasing the costs of food, clothing, and healthcare are keys to unravelling ISIS's popularity and support. (One should be cautioned to not use these *unvalidated* dynamic simulations as definitive quantitative solutions, but only to elucidate cause-and-effect relationships, leverage points, and potential dynamic behaviors over time.)

Many similar simulations are possible using the iThink software.

# Discussion

Linear thinking often leads to unintended consequences and short-term solutions to problems which then recur. This has certainly been the case with America's foreign policy in the Middle East. Systems Thinking has a much greater probability of uncovering the root causes of issues and implementing real solutions.

One linear thinking approach to ISIS is to bomb them back to the Stone Age (Texas GOP Senator Ted Cruz, Kevin Fobbs in *ClashDaily*, Tom Dougherty in *Practical Politicking*, many others). The idea of fighting a military war against this enemy is unlikely to succeed strategically. Even if we succeed in decimating ISIS, the root causes of ISIS-type movements remain. These include corrupt, nepotistic Middle Eastern governments, extensive poverty, an archaic social system in which it is impossible for the disadvantaged to succeed, and disillusioned youth. Eliminating ISIS without addressing the underlying causes will just result in a new ISIS by a different name. Systems Thinking, on the other hand, addresses this issue by examining the underlying causes of ISIS and its appeal. The Systems Thinking solution is to fight not a military war, but a socio-economic one powered by social media.

A second linear thinking concept is that democracy is the best form of government. While this is true in many cases, it is not universally applicable. First of all, the United States does not enjoy a pure democracy, but a *constitutional* democracy in which the people believe in the constitution and the legal system upholds it. In a pure democracy, majority rules and it would be easy for the majority to deny rights to all minorities. Our constitution and Bill of Rights prevent this. The concept that minorities enjoy the same rights as the majority is not understood by many Middle Easterners who view democracy as an opportunity to take control. Secondly, a democratic government works when the people believe in it as the highest law in the land. In a region where religious doctrine trumps secular law, democracy may not succeed. Third, one cannot impose democracy without examining the history of the region. In Iraq, where the majority Shia were repressed for decades by the ruling Bath party, the sudden change to democracy presented an opportunity for revenge. Systems Thinking examines these ramifications and potential unintended consequences and concludes that democracy may not be the best form of government, particularly if it is imposed.

A third linear thinking attitude is that the United States has a right and moral obligation to be the world's policeman. It is clear from our analysis that the use of the "U. S. as World Policemen" mental model will only continue to cause resentment from other countries, increase opposition to U.S. efforts abroad, give rise to support for anti - U.S. activities, and prevent other counties from fighting their own battles. Systems Thinking suggests that the U. S. should encourage ISIS's Middle Eastern neighbors to work together and coordinate efforts to attack and defeat ISIS, with U. S. assistance as advisors and with air support only if requested.

It seems clear that Systems Thinking would yield vastly different approaches to geo-political problems than has linear thinking.

There are different definitions of Systems Thinking (see Monat and Gannon, 2015). Some systems thinking experts believe that systems thinking includes system dynamics modeling, causal loop analysis, and stock-and-flow diagrams, but excludes the Iceberg Model and the patterns and events caused by systemic structures and mental models. One can see from Table 1 above that the inclusion of the Iceberg Model and, specifically, the systemic structures and causative mental models, yield a much richer picture of the systemic behavior patterns of ISIS, the U.S., and other countries.

# Conclusions

We have applied a comprehensive set of Systems Thinking tools (including the Iceberg Model, causal loop diagrams, stock-and-flow diagrams, and dynamic modeling) to analyze ISIS's beliefs, goals, needs, and appeal, and to suggest new strategies for dealing with ISIS. Based on our Iceberg Model analysis, we have also identified numerous mental models and systemic structures that help explain the actions taken by ISIS, the U. S., and other countries. The following summarize the results of our analyses and the conclusions we have reached using a Systems Thinking approach.

*Conclusion 1: The appeal of ISIS is not based upon high-level religious and spiritual ideals, but on basic human needs as articulated by Maslow.* The benefits of the individual System Thinking tools (such as the Iceberg Model) seem clear; however the real power of Systems Thinking becomes apparent when those tools are coupled. By integrating Iceberg Models 1, 5, and 8 from Table 1, one may infer that people join ISIS not out of religious principle but out of the promise of a fairer life with more opportunity in which food, clothing, housing, and medical care are assured. Defeating ISIS will require demonstration that this promise cannot be fulfilled by ISIS and that other approaches *will* fulfill it. The U. S. and its Middle Eastern allies have many more financial resources than does ISIS. If ISIS soldiers are acting on the survival level of Maslow's hierarchy rather than on some high spiritual-imperative level (as some undoubtedly are), then we should offer ISIS soldiers $500/month to fight against their ISIS leaders. This approach worked well for some fighters during the Iraq war. Note that $500/month for 30,000 soldiers comes to just $180,000,000/year, a tiny fraction of our military or foreign affairs budgets. A propaganda campaign targeting ISIS soldiers (and involving social media) should be developed to support this campaign.

*Conclusion 2: The fight against ISIS should not be a military war but a socio-economic one fueled by social media.* This conclusion is justified by Iceberg Models 1, 8, and 9 from Table 1 in conjunction with the system dynamics model depicted in Figures 10-13. The need for exponentially-increasing resources and revenues will be ISIS's undoing; as they cannot continue to provide for the growing number of their people, their people will become disillusioned and defect. The world can accelerate this by denying access to the resources ISIS needs to sustain its populace. Therefore, we should invest time in tracing ISIS's revenue streams and attacking ISIS's infrastructure, such as oil, wheat, ransom for hostages, and taxation/extortion.[1] If ISIS cannot provide food, clothing, healthcare, and a better life for their people, they will lose support. As time moves on, providing these will become harder and harder for ISIS, and they will eventually burn out. It is therefore important to demonstrate that under ISIS, civilian life is NOT better. We should let ISIS grow to control more and more people while continually cutting off their oil, wheat, and ransom revenues; isolate them financially. Allow ISIS to respond by routinely increasing its tax and extortion rates, which will stress their populace and undermine their claim to fairness. Patience and perseverance are essential here. Social media should be used as a weapon against ISIS. Highly respected Sunni muftis should openly challenge ISIS's interpretation of the Koran. A propaganda campaign should demonstrate that life under ISIS is not the path to paradise that is was touted to be and should underscore the suffering of the people under ISIS's control. Individuals should be encouraged to defect and come back to the *true* Sunni religion: the one that will get them to paradise while enjoying life in *this* world.

*Conclusion 3: The U. S. policy in the Middle East must change from forcibly installing democracies to one of an appreciation for the region's culture, history, and politics and one of incenting and re-warding Middle Eastern countries to become less corrupt, less co-ercive, more accountable, and less based on nepotism and hereditary power.* Democracy may not be the best form of government, especially when it is imposed. The problem will not go away until the local Middle Eastern governments become less corrupt, less coercive, and

---

[1] Raqqa's Credit Bank is reputed to be the depository and tax authority for much of ISIS's funds (Hubbard, 2014).

more accountable. Until then, the growing number of unemployed youth in Arab countries will find ISIS and similar concepts attractive because they are more fair, just, and moral, and offer more hope for a good life. *Historically, imposing democracy has not addressed this issue.* Bombing ISIS back to the Stone Age will not solve these problems, and as soon as another cause rises with credible relief on these points, a new ISIS will arise with a different name. These conclusions are based upon Iceberg Model 6 from Table 1 coupled with the causal loop diagrams of Figures 6 and 7.

***Conclusion 4: The U. S. Should Stop Acting as the World's Policeman.*** If one couples Iceberg Models 4, 6, and 7 from Table 1 with the causal loop diagrams depicted in Figures 4, 5, and 7, one may conclude that "The U. S. as World Policeman" feedback loop reinforces the resentment of other countries to U. S. interventions in local conflicts. This policy also leads to the devaluation of freedom and democracy because other countries don't need to fight battles and secure freedom on their own; freedom is devalued when it is simply handed to a populace and not earned. Systems Thinking suggests that instead of fighting ISIS directly, the U. S. should encourage ISIS's Middle Eastern neighbors to coordinate efforts to defeat ISIS, with U. S. advice and air support only if requested.

***Conclusion 5: The greatest threat to world peace is the evolution of Iran as a nuclear Caliphate.*** The causal loop diagrams depicted in Figures 6 and 7 suggest that the recurring cycle of oppression among competing Middle East regimes coupled with U. S. intervention in Iraq contributed to the rise of ISIS as an unintended consequence. The military defeat of ISIS (linear thinking) could also give rise to another unintended consequence, namely the emergence of Iran as another Caliphate with nuclear capability. We know that Iran is pursuing nuclear capabilities; it would not surprise us if ISIS were also seeking weapons-grade uranium. As we postulated earlier, the defeat of ISIS could be an immense threat to the world, and it may be more advantageous to let ISIS continue to keep Iran in check.

# References

Anderson, Virginia, and Johnson, Laura, *Systems Thinking Basics From Concepts to Causal Loops*, Pegasus Communications, Inc., Cambridge, 1997.

Bajekal, Naina, "How to Financially Starve ISIS," *Time*, October 20, 2014.

Bellinger, Gene, "Translating Systems Thinking Diagrams to Stock & Flow Diagrams," 2004, http://www.systems-thinking.org/stsf/stsf.htm

Boardman, John and Sauser, Brian, *Systems Thinking: Coping With 21st Century Problems*, CRC Press, Boca Raton, 2008.

Bronstein, Scott, and Drew Griffin, "Self-Funded and Deep-Rooted: How ISIS Makes its Millions," CNN Investigations, October 7, 2014, http://www.cnn.com/2014/10/06/world/meast/isis-funding/

Dougherty, Tom, *Practical Politicking*, August 20th, 2014, http://practicalpoliticking.com/2014/08/20/its-time-to-bomb-isisisil-into-the-stone-age/

Fobbs, Kevin, *ClashDaily*, August 21st, 2014, http://clashdaily.com/2014/08/isis-beheading-retaliation-solution-bomb-stone-age/

Gladwell, Malcolm. *Blink: The Power Of Thinking Without Thinking*, New York, Little, Brown and Co., 2005.

Goodman, Kemeny and Roberts, "The Language of Systems Thinking: 'Links' and 'Loops'," The Society of Organizational Learning, https://www.solonline.org/?page=Tool_LinksLoops&hhSearchTerms=%22systems+and+thinking%22

Greenfield, Daniel, "Iran's Supreme Ayatollah Bashes US, Calls for Global Caliphate," *Frontpage Magazine*, June 6, 2015.

Hamid, Shadi, "Is There a Method to ISIS's Madness," Brookings Institute, Markaz, http://www.brookings.edu/blogs/markaz/posts/2015/11/24-isis-method-madness-hamid, November 24, 2015.

Hubbard, Ben, "Life in a Jihadist Capital: Order With a Darker Side," July 23, 2014, http://www.nytimes.com/2014/07/24/world/middleeast/islamic-state-controls-raqqa-syria.html

Kauffman, Draper L. Jr., *Systems One: An Introduction to Systems Thinking*, Future Systems Inc., S. A. Carlton, Minneapolis, 1980.

Kim, Daniel H., *Introduction to Systems Thinking*, Pegasus Communications (now Leverage Networks Inc., www.leveragenetworks. com), 1999, ISBN 1-883823-34-X.

Maslow, A.H., "A Theory of Human Motivation," *Psychological Review* **50** (4) 370–96, 1943.

McCants, William, *The ISIS Apocalypse: The History, Strategy, and Doomsday Vision of the Islamic State*, St. Martin Press, New York, 2015.

Meadows, Donella H., *Thinking in Systems: A Primer*, Chelsea Green Publishing, White River Junction, VT, 2008.

Monat, J. and Gannon, T., "What is Systems Thinking? A Review of Selected Literature Plus Recommendations," *Amer. J. of Systems Science* **4** (1), July 2015.

PA Consulting Group, 123 Buckingham Palace Road, London SW1W 9SR, United Kingdom, as presented in the *Times of London*, 28 April 2010.

Peters, Ralph, "The Iranian Dream of a Reborn Persian Empire," *New York Post*, Feb 1, 2015.

"Raqqa, Syria—The New Capital of Isis's Caliphate," *Hot Air*, http://hotair.com/headlines/archives/2014/07/24/raqqa-syria-the-new-capital-of-isiss-caliphate/

Richmond, Barry, *An Introduction to Systems Thinking with iThink*, isee Systems, 2004.

Senge, Peter, *The Fifth Discipline*, Doubleday, New York, 1990; revised 2006.

Senge, Peter, Kleiner, Art, Roberts, Charlotte, Ross, Richard, and Smith, Bryan, *The Fifth Discipline Fieldbook*, Doubleday, New York, 1994.

Thompson, Loren, "Five Reasons The ISIS Fight Isn't About Islam," *Forbes*, February 26, 2015

Von Drehle, David, "The War on ISIS," *Time*, March 9, 2015.

Weinberg, Gerald, *An Introduction to General Systems Thinking*, Dorset House Publishing, New York, 1975, 2001.

Wood, Graeme, "What Isis Really Wants," *The Atlantic*, March 2015, http://www.theatlantic.com/features/archive/2015/02/what-isis-really-wants/384980/

# Explaining Natural Patterns Using Systems Thinking

Jamie P. Monat

Systems Engineering Program
Worcester Polytechnic Institute

*Reprinted from American Journal of Systems Science,* **Vol. 6 No.1, 2018, 1-15**

## Abstract

Patterns in nature are common, from zebra stripes to geese flying in V-formations to the nautilus's spiral. In systems, the presence of a pattern indicates that there are several factors acting in feedback loops; those feedback loops are, in turn, caused by underlying laws or forces such as gravity, electrostatic attraction/repulsion, friction, surface tension, fluid shear, chemical potential, pheromones, and aerodynamic lift. The feedback loops cause the systems to oscillate and the oscillation is interpreted as an emergent pattern. Systems Thinking (specifically the Iceberg Model and causal loops) may be used to explain natural patterns. Understanding what causes natural patterns may help us to influence them, but more importantly, to translate that knowledge to the design and improvement of human-based systems.

## Introduction: Systems Thinking

According to Monat and Gannon (2015a and b, 2017) Systems Thinking is a perspective, a language, and a set of tools. It focuses on relationships among system components (as opposed to the components themselves), it is holistic instead of analytic, it recognizes that systems are dynamic and usually include multiple feedback loops, and it acknowledges that systems often exhibit emergent and self-organizing behaviors. In those previous works, Monat and Gannon have shown how Systems Thinking can explain and address political and socio-economic issues. In this paper, we describe how Systems Thinking can explain a great many natural patterns in the world: why zebras have stripes, why geese fly in a V-formation, why fish school, why the universe looks as it does, and how life began on earth. The explanation begins with the Iceberg Model.

**The Iceberg Model**. Systems Thinking posits that repeated events or objects represent patterns and that those patterns are caused by systemic structure, which is (in turn) caused by underlying forces. The patterns are often *emergent*, meaning that they cannot be predicted from knowledge of the system components; only when those components interact do the patterns emerge. The underlying structures represent the interactions or relationships among system components: the system's stocks, flows, and feedback loops. Structures develop because of natural underlying forces in natural systems or because of mental models in human-designed systems. The Iceberg Model conveniently shows the relationships among events, patterns, structures, and underlying forces (Figure 1).

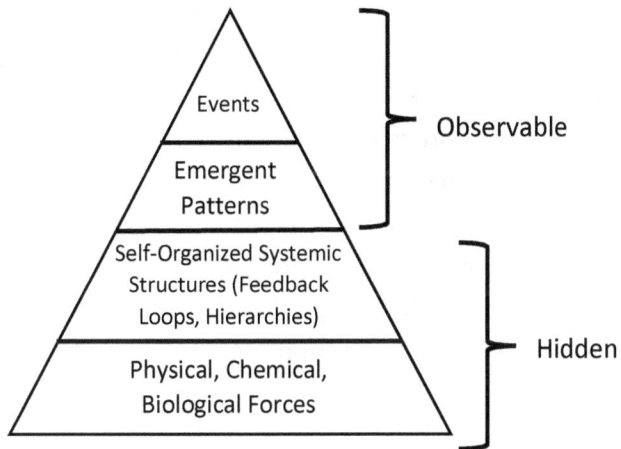

Figure 1. The Iceberg Model for Natural Systems (after Monat and Gannon, 2015a)

For example, the underlying forces of gravity and centrifugal force result in the structure of the solar system, which in turn results in patterns of day and night on planets, the patterns of planetary orbits around the sun, and the patterns of seasons on earth. In a corporation, the underlying mental model that "money motivates employees" results in an incentive compensation structure which, in turn, results in a pattern of excellent performance in some employees. There may be unintended consequences of the structures, as well: an unsavory employee may attempt to take credit for work that she did not do in order to increase her own compensation.

To reiterate, natural patterns (often emergent) are caused by self-organized structures (feedback loops, hierarchies) and their underlying forces. Good systems thinkers try to first recognize and then explain patterns by attempting to understand the underlying structures and forces that yield the patterns. Each of these concepts is explained in greater detail in the following sections.

## Patterns

Businessdictionary.com defines a pattern as a consistent and recurring characteristic or trait that helps in the identification of a phenomenon or problem, and serves as an indicator or model for predicting its future behavior. Patterns may be physical, temporal, behavioral, psychological, or some combination. Examples of physical patterns include stripes on zebras, crystals, sand dunes ripples, compound fly eyes, and termite cathedrals. Temporal patterns are exemplified by predator-prey populations over time, Newton's cradle, my boss being grouchy every Thursday, and the spawning runs of salmon every autumn. Combination physical-temporal patterns include traffic jams, birds flocking, and fish schooling.

## Emergence

Natural patterns are usually *emergent*—that is, they are properties of the system that cannot be predicted from the properties of the system's components. These emergent properties develop as result of the *relationships* among the system elements or between the system elements and the environment. The V-formation of geese, for example, could not be predicted from the flying characteristics of a single goose. It is only when several geese fly together that the V-pattern emerges. Similarly, fish schooling would not be predicted form the swimming characteristics of a single fish. Camazine *et al.* (2001) say, "Emergence refers to a process by which a system of interacting subunits acquires qualitatively new properties that cannot be understood as the simple addition of their individual contributions." Johnson (1999) says,

"Emergence is what happens when an interconnected system of relatively simple elements self-organizes to form more intelligent, more adaptive higher-level behavior. It's a bottom-up model; rather than being engineered by a general or a master planner, emergence begins at the ground level. Systems that at first glance seem vastly different--ant colonies, human brains, cities, immune systems--all turn out to follow the rules of emergence. In each of these systems, agents residing on one scale start producing behavior that lies a scale above them: ants create colonies, urbanites create neighborhoods." Emergent patterns are often a result of self-organized systemic structures such as feedback loops and hierarchies, which are in turn caused by underlying forces.

## Structure

Systemic structure is the way that system elements are linked together and relate to each other, as well as to their environment. Meadows (2008) defines structure as the system's interlocking stocks, flows, and feedback loops. Natural structures <u>are</u> self-organized and include the structure of atoms and solar systems, the physical structure of an animal or crystal, and the herds and flocks of grouped animals. Feedback loops, self-organization, and hierarchies are important elements of natural system structure.

### Feedback Loops

Multiple feedback loops are present in most systems. As shown in Figure 2, these feedback loops may be positive/reinforcing (such as compound interest on a savings account) or negative/balancing (as a home thermostat or cruise control on an automobile.)

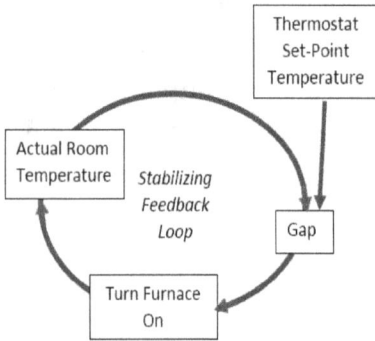

A Balancing (-) Feedback Loop:       A Reinforcing (+) Feedback Loop:
   Home Thermostat in Winter                    Savings Account

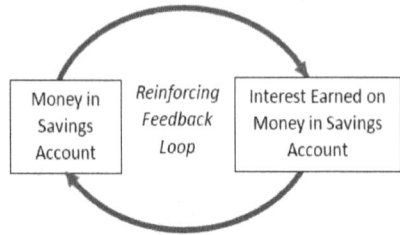

Figure 2. Balancing and Reinforcing Feedback Loops

Feedback loops cause oscillations in several different ways. Stabilizing (balancing) Feedback Loops with delays cause oscillation because of the delays. (Delays arise from several sources: physical delays may exist due to inertia or momentum, physiological delays may exist because of finite gestation periods or reaction times, perceptual delays exist because of the time required for plants and animals to realize and react to situations.) Figure 3 shows a Behavior-Over-Time plot for a predator-prey system involving rabbits and foxes.

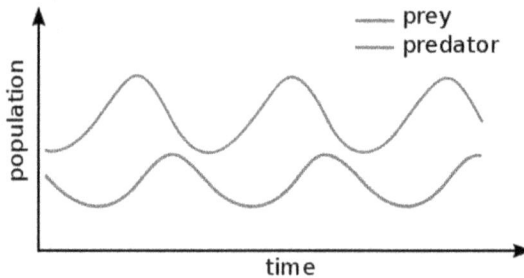

Figure 3. Behavior-Over-Time Plot for Predator and Prey

Suppose that initially there is a population of rabbits living happily in a field at the edge of a woods. They mostly eat, sleep, and procreate. One day, a pair of foxes wander buy and decide to settle down because of the abundant cunicular food supply. The foxes prey on the rabbits and raise a healthy litter of fox kits, who, in turn, grow and raise their own families, all of which prey on the rabbits. Eventually, there are so many foxes that the population of rabbits starts to decline due to predation. Fewer rabbits means that fewer foxes can be supported, but the effect is not immediate: it may take months, or even years, for the fox population to decline in response to the declining rabbit population. This delay is because of the finite gestation period for both rabbits and foxes; the finite time required to hunt, kill, and digest a rabbit; and for the time required for the lack of food to affect the foxes' decision to move on. Eventually, it does, the rabbit population starts to rebound, the foxes proliferate in response, and the cycle repeats. If there were no delays, the system would not oscillate, but instead would reach a steady-state value and stay there.

Figure 4's behavior-over-time plot (which corresponds to the home thermostat causal loop diagram of Figure 2) shows another example of systemic physical delays causing oscillations in a domestic heating system.

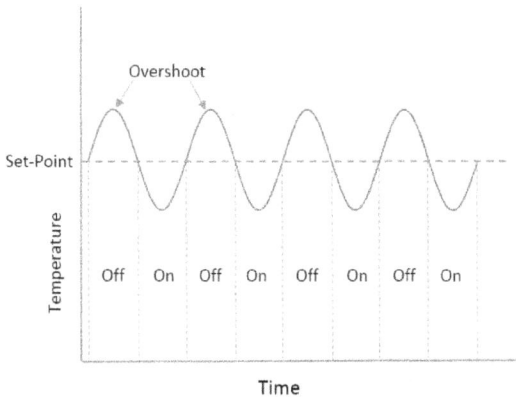

Figure 4. Temperature Oscillation in a Domestic Heating System

When it is cold out and the outside temperature drops below a set-point, the thermostat causes the home's furnace to start up. But heat transfer takes time, and because of the thermal mass of the house, the temperature response to increases in heat is not immediate. Furthermore, once the set-point temperature is reached and the furnace shuts down, the temperature continues to rise a little as residual heat in the system's pipes and fluids is transferred to the home even though the burner is off; the result is an overshoot of the set-point temperature. This yields the sinusoidal temperature oscillation around the set-point depicted in Figure 4. Note that if there were no delays, the system temperature would approach its set-point smoothly and without oscillation.

Positive (reinforcing) feedback loops also cause oscillations when the reinforcing feedback loop is broken due to physical constraints. Examples include sand dune ripples and ocean waves for which gravity collapses the mounding piles of sand or water that are developed due to reinforcing feedback loops; and some cloud formations for which the cloud-forming reinforcing feedback loop is broken by turbulent mixing. (These patterns will be described more fully subsequently.)

So both stabilizing feedback loops with delays and reinforcing feedback loops with breaks cause oscillations. Those oscillations represent patterns. (Note that system dynamic software such as isee's *Stella Architect* and Ventana System's *VenSim* may be used to develop dynamic models that show the oscillatory behavior.) The natural feedback loops often develop because of self-organization.

*Self-Organization*
Natural systems tend to self-organize into structures due to underlying forces. Packs of wolves develop an organizational hierarchy. Ants organize into marching columns. Wildebeests organize into herds. Even inanimate systems self-organize. As a super-saturated salt solution cools, Na and Cl atoms organize into a precisely-arranged crystal. Groups of stars organize into spiral or globular clusters. DNA components self-assemble into DNA molecules, then disassemble and replicate. Camazine (2001) says, "Self-organization is a process in which a pattern at the global level of a system emerges solely from numerous interactions among the lower-level components of the system. Moreover, the rules specifying interactions among the system's components are executed using only local information, without reference to the global pattern." According to Beckenkamp (2006), "Self-organization exists if – independent of the intentions or even existence of an organizer or a central plan –

regular or arranged patterns emerge from the interactions in the system itself." Self-organizing activities do not happen by chance, but instead due to specific underlying forces. Natural hierarchies are a common structure resulting from self-organization.

*Hierarchies*

Many complex natural systems are hierarchical—that is, they comprise sub-systems which, in turn, comprise sub-systems of their own, etc., until the most fundamental sub-system comprises the raw system elements. (Systems Engineers would call this concept "System of Systems" or "SoS.") Examples include the universe (galactic clusters, galaxies, solar systems, planets with moons); the human body (the human organism; the circulatory, reproductive, digestive, and other systems; the blood vessels and organs; the organ sub-structures; the individual cells); ecosystems (plants and animals, trees, tree components such as leaves, leaf sub-structures such as veins and epidermis, cells, cell components,) and many load-bearing biological tissues such as wood and bone (Fratzl and Weinkamer, 2007.) Natural hierarchies persist because of benefits to both the lower levels and upper levels in the structure: each sub-unit benefits from the organization, protection, and support of the higher level; and the higher level benefits from the added functionality and quantity of the sub-unit. Hierarchies thus represent symbiotic relationships between sub-units and their superiors, in a chain of reinforcing feedback loops that permeates the entire system.

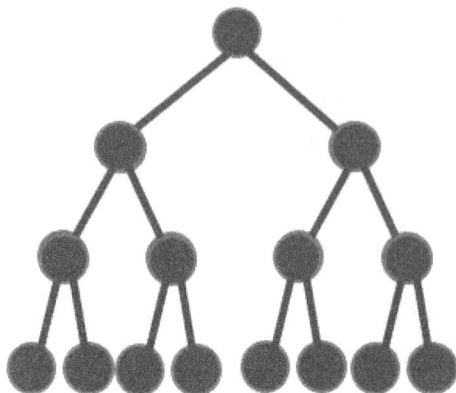

Figure 5. Hierarchical Pattern

Hierarchies themselves are often patterns (Figure 5). For example, the unit of a mass orbiting around a larger, central mass is repeated at several different scales in the universe. But it may be hard to see the hierarchical structure in complex systems. Recognizing the hierarchical structure and then understanding how that structure evolved over time helps understand how the system came to be.

To help with this, one must understand that self-organizing hierarchies evolve from the bottom up: multi-cell organisms from single-cell; complex protein configurations from simpler ones; a complex human brain from a primitive collection of ganglia. The underlying forces driving this self-organization may be chemical, physical, electrical, or other; but the newly self-organized structure typically has a competitive advantage, which leads to its evolutionary persistence. An example is the evolution of herding behavior in wildebeests. Individual wildebeests were easy picking for predators. By chance or by some natural affinity, several wildebeests grouped together and were safer from predators. In subsequent generations, those wildebeests that displayed this herding tendency survived better than singular animals, so the tendency to herd was passed down. Thus a hierarchy was formed: individual wildebeests into groups; groups into herds. The reinforcing feedback loop in this situation is clear: individual wildebeests with superior survival traits survive better than those who don't and therefore more of them pass down their genes. Their progeny survive better and, in turn, more of them pass down their genes, etc. Eventually, those wildebeests with better survival traits or "fitness" predominate and those with poorer fitness die out. This is the basis for natural selection and evolution.

Meadows (2008) advances another reason for the prevalence and stability of natural hierarchies. She states that "Complex systems can evolve from simple systems only if there are stable intermediate forms. The resulting complex forms will naturally be hierarchic. That may explain why hierarchies are so common in the systems nature presents to us. Among all possible complex forms, hierarchies are the only ones that have had time to evolve."

The modern prokaryotic plant cell is a good example of natural hierarchies facilitating the evolution of complex systems. Modern plant cells can efficiently photosynthesize and transfer energy, but this was not the case for their primitive ancestors. It is believed that modern plant cells evolved from primitive eukaryotes by engulfing specialized independent bacteria such as aerobic prokaryotes (which evolved into mitochondria, the cell's energy-transfer organelle) and photosynthetic prokaryotes (which evolved into chloroplasts, the cell's photosynthetic organelle) to become complex photosynthetic eukaryotes; *viz.* modern plant cells (Scitable, 2017.) This process (termed endosymbiosis) was itself a feedback loop: the parent cell benefits from the energy production and photosynthesis of the mitochondria and chloroplasts; and they in turn benefit from the protection and nutrients of the parent cell. This evolution would have taken much longer, and perhaps not have happened at all, were it not for the availability of the 3 primitive sub-units: the primitive eukaryote, the aerobic bacterium, and the photosynthetic bacterium, all lower levels on the hierarchy.

Much natural self-organization into hierarchies works similarly. It is much easier to understand the structures of DNA, complex life forms, and the universe if one understands that the end result evolved from simpler sub-systems that were themselves more structured than their raw components. The hierarchical organization of many natural structures contributes to the patterns we see while explaining how complex systems evolved.

**Underlying Forces (Mechanisms of Self-Organization)**
In natural systems, there are many underlying forces yielding self-organization:

- Physical mechanisms such as gravity, electromagnetic force, the strong and weak nuclear forces, aerodynamic lift, fluid shear, friction, pressure, mechanical forces, centrifugal force
- Chemical mechanisms such as pheromones, the electronic charge of ions, Vander Walls forces, hydrophilicity/ hydrophobicity, chemical potential, the physical shape and structure of molecules
- Instinctive survival mechanisms such as swarming, flocking, schooling, herding to afford a competitive advantage
- Comfort/Discomfort perceptions such as safety zones, sensitivity to heterogeneity, separation, alignment, and cohesion

In human-designed systems, the underlying forces are typically mental models such as "incentive compensation increases productivity," "honesty is the best policy," and "competition brings out the best in people." Several examples of patterns, their underlying structures, and the structures' causative underlying forces are presented in Table I.

Table I. Some Examples of Patterns, their Causative Structures, and their Underlying Forces.

| Underlying Forces | Self-Organized Structure | Emergent Pattern |
|---|---|---|
| Natural selection; Survival of the Fittest; Evolution | The Human Brain | Self-Awareness |

| | | |
|---|---|---|
| Aerodynamic Lift, Minimization of energy expended | Rules: Fly behind and to the side of the preceding goose to where it's easiest to stay aloft –but not too close! (feedback loops) | V-Formation |
| Perception of safety and need to belong (Mental Models/ Instincts of birds and fish) | Rules: Maintain safe distance, average direction, average velocity (feedback loops) | Murmuration, flocking, schooling |
| Mental Model: We can grow our food in 1 place instead of following herds | Organized groups of humans sharing permanent common geography | Countries |
| I need to get to my destination as fast as possible without an accident (Mental Model) | Rules: Safe following distance and speed (feedback loops) | Traffic Jam Patterns |
| Gravity and Centrifugal Force; insolation | Planetary Motion: orbits, tilt of the earth | Hibernation; Dormancy |
| Survival | Ostracization of people Not Like Me | Bigotry, Racism, Prejudice |

**Patterns in Nature**

There are hundreds of patterns occurring naturally. Although some of these seem beyond understanding, systems thinking may be used to explain each of them.

*Sand Dune Ripples.* The intricate ripples in sand dunes (Figure 6) are explainable using Systems Thinking. As wind moves along a flat bed

Figure 6. Sand Dune Ripples

of sand, it may happen to dislodge a single sand grain, carry it along for a bit, and then deposit it on top of the flat sand surface. This obstacle causes the horizontal wind to divert upwards. This upward wind direction then carries additional sand grains over the original grain and deposits them. The pile of sand is now a little bigger and causes the wind to bend upwards even more. As the process continues, more and more sand accumulates until the slope is so steep that a mini-avalanche of sand occurs due to gravity. Meanwhile, the shape of the newly-formed ripple causes the wind to detach from the surface, re-attach downstream, and repeat the process (Makse, 2017.)

In this situation, the feedback loop is physical: The result of the wind force (the sand ripple) impacts the wind force itself (the wind direction) which then impacts the shape of the ripple, which, in turn, impacts the direction of the wind in a reinforcing feedback loop that continues until gravity topples the top of the ripple. Wind accumulates; gravity topples. This is an example of a reinforcing feedback loop proceeding to failure (due to an interruption by gravity) and then starting up again.

*Crystals.* Systems thinking may be used to explain crystal formation using sodium chloride (NaCl) as an example (Figure 7). In an aqueous solution, the positively charged sodium atoms $Na^+$ and negatively

Figure 7. Salt Crystals

charged chloride atoms Cl⁻ are fully dissociated as charged ions. If the solution cools, the attractive electrostatic forces will cause a Na⁺ and a Cl⁻ atom to join together to form a NaCl molecule, the sodium end of which has a + charge and the chloride end of which has a negative charge. Once this initial "seed" molecule forms, its electrostatic forces

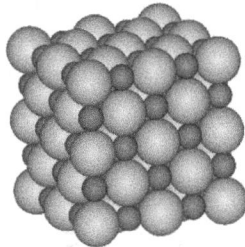

Figure 8. NaCl Ion Pattern. The green spheres represent Cl⁻ ions and the purple spheres represent Na⁺ ions.

will attract other Na⁺ and Cl⁻ ions from the solution: Na⁺ will bind to Cl⁻ and vice versa. This will occur in 3 dimensions (Figure 8), meaning that the structure will grow in all directions, yielding a sold-state crystal which precipitates out of the solution. It will be highly ordered in that all Cl atoms will be surrounded by Na atoms, and all Na atoms will be surrounded by Cl- atoms, as shown in Figure 8. The feedback loop is evident: + attracts – and – attracts + in a reinforcing loop which causes the crystal to grow. The underlying forces are electrostatic attraction and repulsion. Growth will continue until free Na and Cl ions are no longer available.

*Newton's Cradle.* Newton's cradle is a pendulum-like device in which multiple pendulum bobs transfer energy and swing in mesmerizing patterns (Figure 9). As for a simple pendulum, the magnitude of the

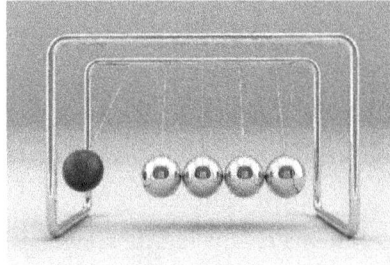

Figure 9. Newton's Cradle

restoring force (which is the tangential component of gravitational force) increases the farther the bob is displaced from its equilibrium position. But the bob position, in turn, is a function of the forces acting on it. Adding a few more bobs and efficient momentum transfer yields the Newton's Cradle patterns. The feedback loop is shown in Figure 10.

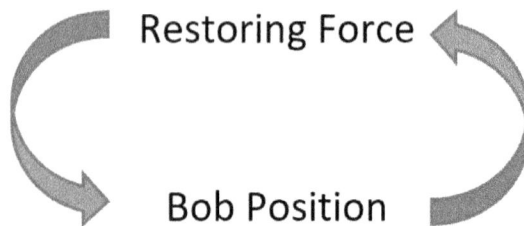

## Restoring Force

## Bob Position

Figure 10. Stabilizing Feedback Loop for Newton's Cradle

The restoring force impacts the bob's position and the bob's position impacts the restoring force. Because the restoring force resists bob positions further from equilibrium (after all, it is a "restoring" force) this feedback loop is negative, or stabilizing.

*Zebra Stripes.* Scientists have studied the causes of zebra stripes (Figure 11) for a long time. Current thinking invokes a balancing

Figure 11. Zebra Stripes

feedback process involving an "activator" chemical that turns on the dark pigment melanin, and an "inhibitor" that suppresses the activator. These 2 chemicals impact each other in a negative feedback loop:

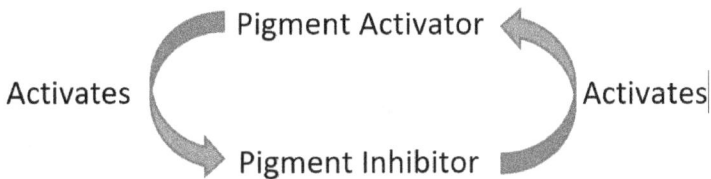

Figure 12. Balancing Feedback Loop for Zebra Stripes

The result is a chemical oscillation yielding the stripes that we see. This phenomenon was modeled and described by Alan Turing (1952) who studied animal patterns and developed a "reaction-diffusion" mathematical model in which a rapidly diffusing inhibitor reacts with a slowly diffusing activator. The Turing Reaction-Diffusion model involves several adjustable variables that control diffusion and reaction rates. By adjusting these variables, scientists have been able to create a variety of animal coloration patterns (Howard Hughes Medical Institute, 2005.) The zebra stripes represent oscillations, which are caused by feedback loops.

*Compound Eyes.* Many insect eyes display beautiful patterns with hundreds of individual lenses laid out in geodesic-like hemispheres (see Figure 13). Myers (2006) says, "You might even say, if you found

Figure 13. Fly Eyes

a fly eye upon a heath, that you could not imagine how something so beautiful and perfectly arranged and organized could possibly have come into existence without some superhuman engineering." He explains this as follows: "This is easy to understand: the cells in the fly eye are doing *The Wave.* After the wave has passed, cells form small clusters and signal each other; one cell sets itself apart and begins to differentiate into the R8 cell, and recruits two neighbors to form the R2 and R5 cells. Subsequently, R3 and R4 are drawn in, then R1 and R6, and finally, R7. The cells all have the same orientation to one another. It's all quite mechanical and reliable, mediated by a small set of genes...... EGF-R signalling is activated by the ligand, Spitz, and inhibited by the secreted protein, Argos." Without understanding the specific cell nomenclature and biomechanics, it is still clear that the ligand Spitz and the protein Argos are opposing forces that impact each other in a feedback loop. This is similar to the zebra stripe activator-inhibitor mechanism, but causing physical structure instead of pigmentation.

*Geese V-Formations.* If you had never seen geese in flight, even if you understood the lift generated by trailing vortices, you would never

Figure 14. Geese in V-Formation

guess that when you put a flock of geese together, they will form a V (Figure 14). The pattern emerges because of the relationships among the components of the system. When an airfoil (such as a goose's wing) moves through the air, a vortex trails off the airfoil's tip. The vortex is a spiral of air rotating about a horizontal axis (see Figure 15)

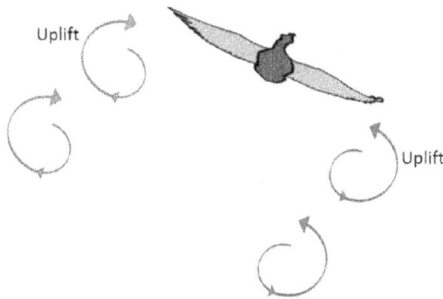

Figure 15. Trailing Vortices

with one side of the vortex moving up and the other side moving down. The upward-moving side provides additional lift and makes it easier for another object to fly at that location. So a trailing goose naturally moves around behind a leading goose until it finds the greatest lift. The negative (or stabilizing) feedback loop in this case is the reduced lift as the goose flies further away from the center of the rising side of the vortex. Goose flies away from the uplift side of the vortex→less lift→goose flies closer toward the uplift side of the vortex→more lift. The results when several geese fly together is the iconic V. The oscillation in this case is more subtle. It occurs because a single goose following a preceding goose's trailing vortex becomes the leader for a following goose. Thus every goose (except the lead and last geese in a gaggle) is both a cause and effect, as shown in Figure 16.

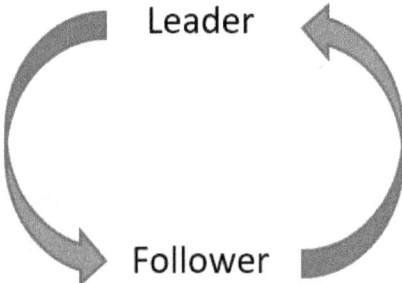

Figure 16. Reinforcing Feedback Loop for Geese

The V-formation may be interpreted as a broken (but still functional!) reinforcing feedback loop. Imagine a gaggle of geese flying in a circle, one behind the other and offset, each taking advantage of its

Figure 17. Geese Flying in a Circle

predecessor's trailing vortex (Figure 17). This would be a very efficient flight structure with the only disadvantage being that the geese would not fly anywhere. If the circle is broken, one obtains the classic V-formation. Deneubourg (1989, 1990) has studied and written extensively on pattern formation in social animals such as ruminant herds and fish.

*Fish Schools.* Fish schooling (Figure 18) may be explained using systems thinking at 2 different scales, both of which are evolutionary

Figure 18 Fish School

and related to survival. At the macro level, there is safety in numbers; being with a group of your species reduces the likelihood that an individual will be preyed upon. Schooling was reinforced as a greater number of fish who tended to school together survived to pass their genes on to subsequent generations, while those who did not school died off as they were preyed upon. The reinforcing feedback loop strengthened the schooling tendency in every new generation. At the micro level, the fish follow the 3 rules of group behavior: maintain the general speed of your neighbors, maintain the general direction of your neighbors, and don't get too close to your neighbors. Note that these rules form several counteracting feedback loops: if I go too slow, speed up; if I steer too far left, correct by steering right, and if I get too close, move farther away. But the dominant feedback loop, which applies to many animal group movements, is depicted in Figure 19: The position and motion of the group impacts the position and motion of each individual which, in turn, collectively impacts the position and motion of the group.

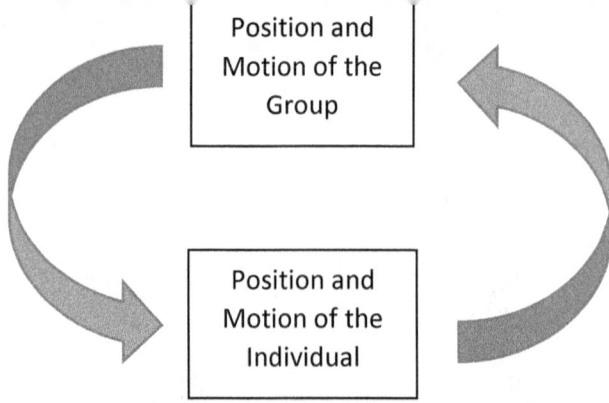

Figure 19. The Dominant Feedback Loop Describing Group Motion

The result is that the fish school appears to oscillate or undulate as it moves through the water (Muhammad Saif Ullah et al., 2016; Miller et al., 2017.)

*Herd Patterns.* Herding behavior describes the tendency of groups of animals to move together in a pattern. It is observed in wolves, deer, sheep, goats, horses, bison, cattle, wildebeests, caribou, elephants, and many other species. The behavior often looks organized as depicted in

Figure 20. Arctic Caribou Herd

Figure 20. This structure derives from 2 instincts: 1) it is safer to stick with a group than to be out alone, 2) avoid getting trampled. Similar to fish schooling, these instincts likely evolved as survival mechanisms—those who happened to follow these "rules" survived better than those who did not, and were thus able to pass on their genes to their offspring so that these instincts were perpetuated and strengthened in subsequent generations. The instincts resulted in 3 behavioral "rules:" 1) follow the general direction of your neighbors, 2) maintain the general speed of your neighbors, and 3) don't get too close to your neighbors. Note the feedback loops established by these rules, which yield the familiar undulating herd patterns exemplified in Figure 20 deriving from the generic feedback loop shown in Figure 19. It is interesting to note that these stabilizing feedback loops may yield relaxed, slow movement or stampedes, and anything in between. Unlike for birds or fish, the oscillatory nature of land animal group behavior is limited to 2 dimensions.

*Bird Murmuration.* Bird murmuration is a spectacular natural phenomenon, characterized as a shape-shifting cloud of thousands of birds whose movements appear

Figure 21. Examples of Bird Murmuration

coordinated (Figure 21). As for schooling and herding animals, the natural forces underlying murmuration are instinctive, resulting in 3 familiar "group movement" rules: 1) *alignment*: follow the general direction of your flockmates, 2) *cohesion*: steer toward the average position of your flockmates, and 3) *separation*: maintain a "comfort zone"-- don't get too close to your flockmates (Reynolds, 1987 and 2017.) Adding a little randomness caused by wind or distractions yields the mesmerizing aerial clouds that sometimes appear to be a single living, moving creature. The simple instinctive rules (mental models) yield balancing feedback loops at the micro scale (e.g. "I am too far from my neighbors: fly closer" or "I am too close to my right-hand neighbor; fly left") and the overarching feedback loop shown in Figure 19, resulting in the concomitant aerial oscillations. Craig Reynolds and other researchers have used the 3 group movement rules to create convincing computer simulations of murmuration (Reynolds, 2017.)

*Black Holes.* One of the more interesting natural patterns is that of a black hole (Figure 22), where a reinforcing feedback loop generates a natural singularity. Cosmologists estimate that there are billions of

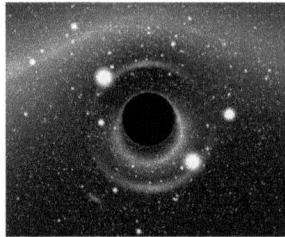

Figure 22. Black Hole

Black holes in the universe. The black hole displays radial symmetry in one plane and the singularity in the perpendicular plane. The feedback loop is caused by the interactions of gravity and mass: gravity draws in more mass, the increased mass increases the gravitational force (spatial distortion), which (in turn) draws in even more mass, as shown in Figure 23.

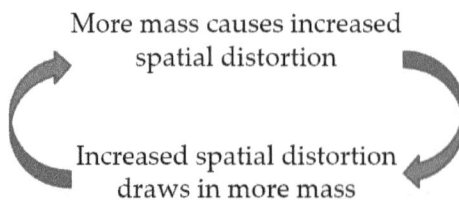

More mass causes increased spatial distortion

Increased spatial distortion draws in more mass

Figure 23. The Black Hole Reinforcing Feedback Loop

This is an example of a natural reinforcing feedback loop run to the extreme. Smolin (1999) speculates that each black hole contains a new universe that may generate new black holes in a never-ending reinforcing feedback loop and pattern of new universe generation.

*Kelvin-Helmholz Clouds*. These clouds (Figure 24) form when 2 stratified air layers of different densities move horizontally at different relative velocities, creating shear at the interface. The dryer, faster-

Figure 24. Kelvin-Helmholz Clouds

moving upper layer typically scoops up pieces of the lower, denser cloud layer and captures them in a vortex, as shown in Figure 25 (after Smyth and Moum, 2012.)

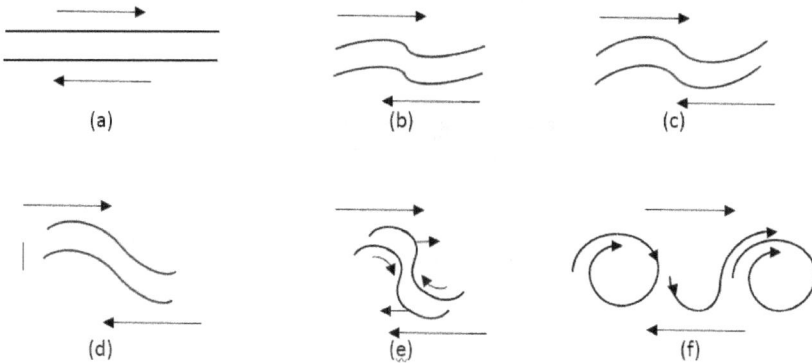

Figure 25. The Formation of Kelvin-Helmholz Clouds

The vortex develops into a cylindrical (horizontal axis) rotating cloud (reminiscent of a breaking ocean wave), but it dissipates due to turbulent mixing and evaporation; and another one forms downstream. Some believe that these clouds were the inspiration for Van Gogh's "Starry Night." The feedback loop in this case (Figure 26) is reinforcing:

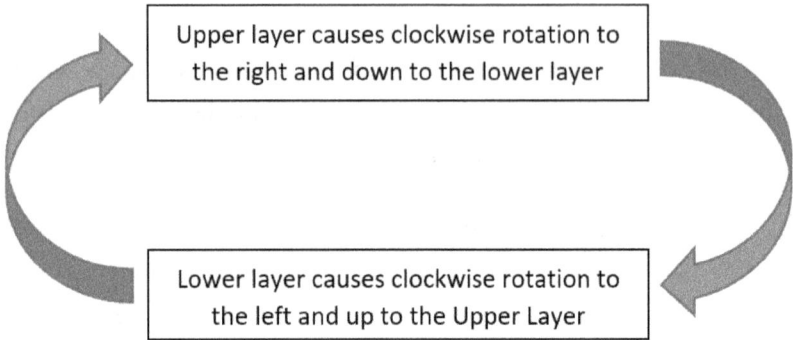

| Upper layer causes clockwise rotation to the right and down to the lower layer |
| Lower layer causes clockwise rotation to the left and up to the Upper Layer |

Figure 26. Reinforcing Feedback Loop for Kelvin-Helmholz Cloud Formation

This is another example of a reinforcing feedback loop that is interrupted (this time by turbulence and mixing), to start again.
*Ocean waves* (Figure 27) follow a similar mechanism, except that the 2 fluid layers are air and water, and because of the higher density of water, gravity plays a bigger role in causing the waves to break, thus

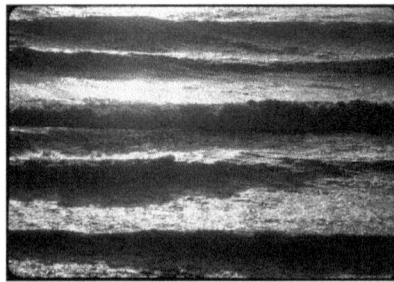

Figure 27. Pattern of Ocean Waves in a Line

interrupting the reinforcing feedback loop that is attempting to create vortices.

Space does not permit a detailed explanation of every natural pattern; however Figure 28 shows some more that may also be understood through the application of Systems Thinking.

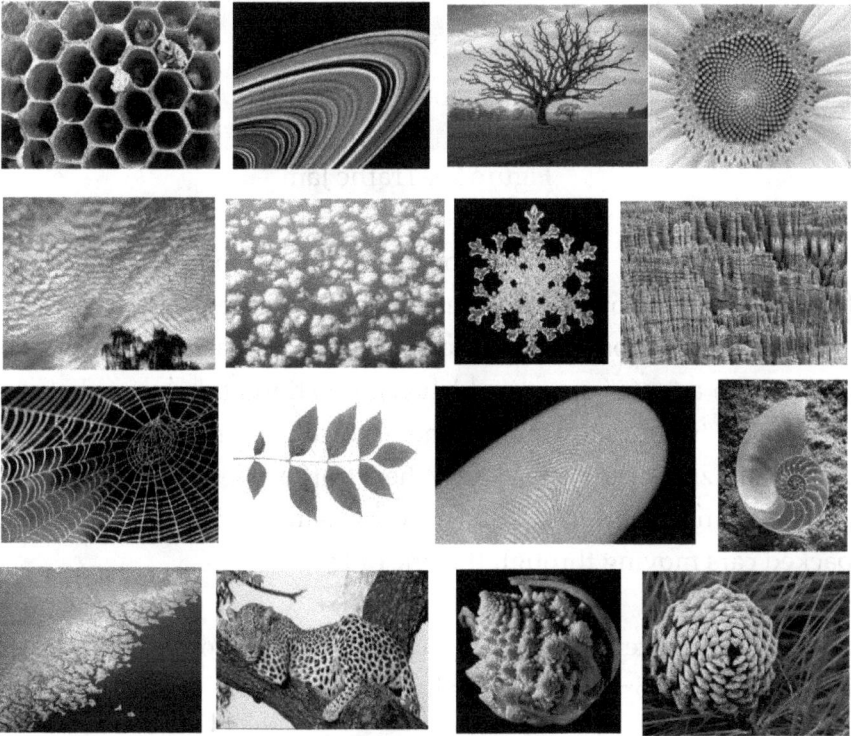

Figure 28. Additional Natural Patterns

## Patterns Deriving from Man's Activities

Although this paper focuses on natural patterns, there are many patterns that derive from man's activities. Traffic patterns (especially traffic jams; Figure 29) are 1-dimensional versions of the herding/

Figure 29. Traffic Jam

schooling/flocking behavior observed in animals. Drivers are following 2 basic opposing rules: 1) get to my destination as fast as possible and 2) don't have an accident. These basic mental models translate to a "safe" speed (which is a function of the space between cars) and a "safe" following distance (which is a function of the speed) in a stabilizing feedback loop. When traffic piles up in a jam, the pattern from above appears as a backwards-travelling wave of denser-packed cars moving through the line of traffic.

The network pattern deriving from American city location (Figure 30) is also based on opposing forces in a feedback loop.

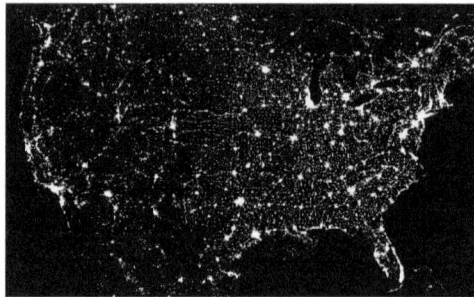
Figure 30. Pattern of U. S. Cities

One force that drives cities to form is the availability of human and natural resources, good transportation, and an attractive business and living environment. Competition for those resources limits the size of cities and enforces a degree of separation.

## The Structure of the Universe

There are several repeated patterns in the universe, from the pinwheel shape of spiral galaxies to the patterns of planets orbiting their sun and moons orbiting their planets. Systems Thinking teaches us that patterns are caused by underlying structure, which (in turn) is caused by underlying forces. These concepts are useful in understanding those patterns.

The universe appears to be hierarchic: planets (some with moons) embedded within solar systems, solar systems within galaxies, and galaxies in clusters, and these basic structures are repeated many times. Can Systems Thinking and the underlying forces of gravity, electromagnetism, centrifugal, and the nuclear forces explain the structure of the universe, and is the universe self-organizing?

Figure 31. Spiral Galaxy NGC 4414

As an example, consider the repeating pinwheel pattern of spiral galaxies in which the shape of each curved radial arm is repeated many times (Figure 31). This pattern is formed by 2 opposing forces: centripetal force $F_c$ ($F_c = mv^2/r$) and gravity $F_g$ ($F_g = GMm/r^2$) where m=the mass of a galactic star, v=the tangential velocity of the star, G= the universal gravitational constant, M = the mass of the galaxy (concentrated at its center), and r= the radial distance of the star from the center of mass. $F_g$ and $F_c$ must be balanced, or else the stars would either all collapse into the galactic center or fly out of orbit and leave the galaxy. If one equates these 2 forces, one obtains the equation $v=(GM)^{.5}/r^{.5}$ which indicates that the tangential velocity of a star in the galaxy is proportional to the square root of the reciprocal of the star's distance from the center of the galaxy: that is, more distant stars revolve more slowly[2]. Over time, the more distant stars lag behind the closer stars in their counter-clockwise motion around the galaxy's center of gravity. The result is the classic pinwheel shape shown in Figure 31.

This same analysis applies to solar systems: more distant planets revolve more slowly around

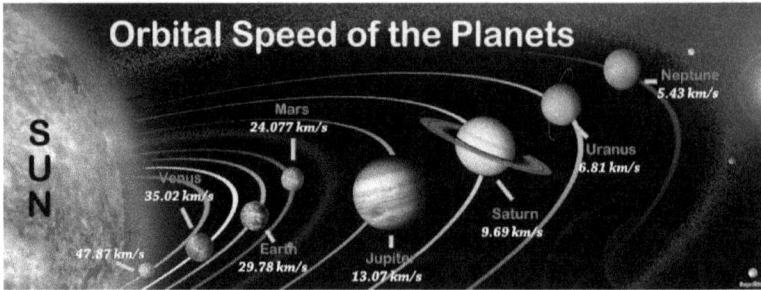
Figure 32. Orbital Speeds of Planets

the sun (Figure 32). It also applies to moons circling a planet. Jupiter's moons, for example, precisely follow the inverse square-root law of orbital velocity predicted by Keplerian mechanics.

These self-organized patterns are thus the result of gravity ($F_g$) and centrifugal ($F_c$) forces acting in opposition. The underlying forces establish a stabilizing feedback loop that keeps them balanced: if, due to some random perturbation, the orbiting body moves closer to (or farther from) its central mass, its orbital velocity will increase (or decrease) so that $F_c = F_g$ as shown in Figure 33.

---

[2]Present-day studies indicate that the tangential velocity of many galactic stars do not follow the inverse square-root law of Keplerian mechanics. This observation has led to the speculation of both the existence of dark matter and a modification to Newton's law of universal gravitation (Milgrom, 2015). However, all known theories account for the outer galactic stars moving slower than inner stars, although they vary in details.

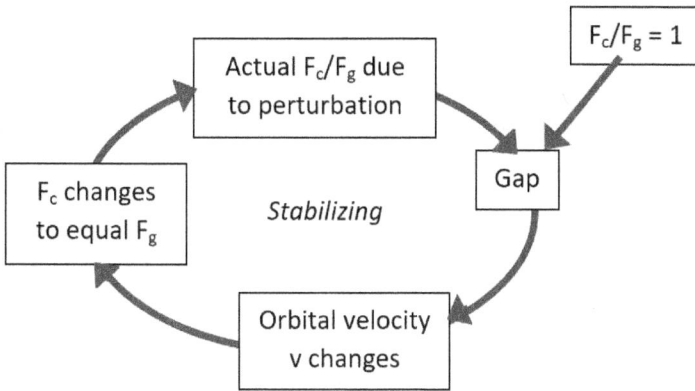

Figure 33. Stabilizing Feedback Loop of Orbital Mechanics

But why do spiral galaxies display an azimuthal oscillation of star density structured as several curved arms extending from the nucleus? Each arm is believed to be a region of concentrated new star formation with a greater density of bigger, brighter stars than the inter-arm regions. Gerola and Seiden (1978) believe that stars self-propagate in a reinforcing feedback loop. When a massive star explodes in a supernova, the resulting shock wave triggers star formation in the nearby stellar medium. The newly formed stars live out their lives and also eventually explode in supernovae, propagating more new stars in a chain reaction. But after each star explodes, the local region is too hot for the interstellar gas to condense to form new stars. So each region alternatively experiences new star formation followed by a period of quiescence, until it cools sufficiently to again allow new star formation. If one were to view the galaxy from above over billions of years, one would see various regions blinking on and off as bright regions seem to ignite their neighbors and then go dark. Gerola and Seiden have shown that due to the differential rotation of the galaxy, each star-forming region stretches into a curve. Over time, these brighter curved regions seem to coalesce into spiral arms.

This reinforcing feedback loop of star formation is supported by a closely-related stabilizing feedback loop: interstellar gas and dust condense (due to gravity) to form stars, which (in turn) eventually explode returning their gas and dust to the interstellar medium, where it again can condense into stars. Thus galaxy formation, shape, and stability are controlled by both reinforcing and stabilizing feedback loops acting in concert.

Lee Smolin (1999) hypothesizes a much grander vision of how feedback loops and self-organization impact the structure of our universe; one in which there have been billions of universes, each with different physical laws, some of which were conducive to life and some of which were not. He argues that the same conditions that were conducive to life also generated many stars, which in turn generated many black holes, each of which represented a new universe. In his model, the star-forming universes thus generated many more star-forming universes in a reinforcing feedback loop, while the star-depleted universes died out. Since each new universe has a slightly different set of physical laws, eventually, there was an extremely high probability that a universe materialized that would be full of both stars and life. This vision suggests that the laws of physics (and their concomitant universes) are not fixed, but evolve over time in a form of Darwinian natural selection to a life-supporting universe like the one we inhabit.

Smolin says, "It then seems that our life is situated inside a nested hierarchy of self-organized systems that begin with our local ecologies and extend upwards at least to the galaxy. Each of these levels are non-equilibrium systems (open and active) that owe their existence to processes of self-organization, that are in turn driven by cycles of energy and materials in the level about them. It is then tempting to ask if this relationship extends further than the galaxy. ... Must there be a non-equilibrium system (open and active) in which sits our galaxy? Is there a sense in which the universe as a whole could be a non-equilibrium, self-organized system?"

These intriguing Systems Thinking theories suggest that naturally occurring feedback loops and their underlying forces yielded a universe that is self-organized, hierarchical, and full of the structures and patterns that we observe.

## The Origins of Life on Earth

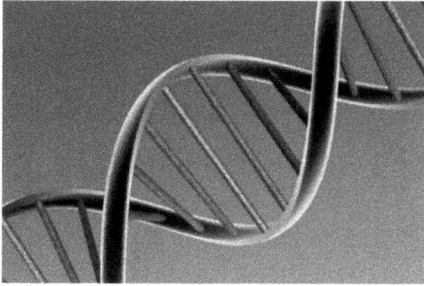

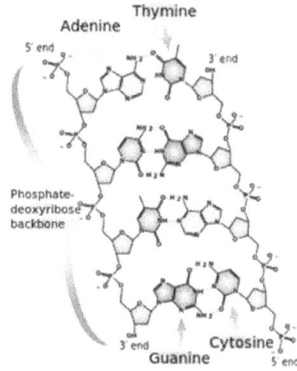

Figure 34. Pattern of DNA

DNA forms a beautiful double helix molecular pattern (Figure 34) with a basic phosphate-ribose-base structure repeating millions of times. Could Systems Thinking concepts, specifically hierarchical system evolution and molecular self-organization, explain the origins of life on earth? Conventional wisdom argues that a DNA-like precursor was the first form of reproductive life. Sir Fred Hoyle and Chandra Wickramasinghe (1981) have demonstrated the statistical improbability of all the elements in the Archean environment *randomly* coming together in just the right sequence to form a molecule of DNA (which comprises millions of atoms.) But was it truly random? Inspection reveals that the DNA molecule is *hierarchic*, with 3 repeating fairly simple sub-units: a phosphate group, a ribose (sugar) group, and a base (adenine, cytosine, thymine, or guanine) connected in a chain and mated to a complementary chain in a double helix. Could a complex DNA molecule have evolved from these much simpler sub-units? The probability of DNA self-organization increases dramatically if its evolution is split into stages: 1) self-organization of each of the 3 basic sub-units independently followed by 2) self-organization of those sub-units into a section of DNA, and then 3) the linking of many such sections to form a long molecule. Scientists have been able to demonstrate the self-organization of amino acids, the

271

sugar, the phosphate, and some of the base groups from a primordial soup of base chemicals (Miller and Urey (1953 and 1959), Oro (1960 and 1961), Keller, Turchyn, & Ralser (2014)), but they have not yet succeeded in demonstrating the self-organization of those constituents into a nucleotide. But self-replication of inanimate chemical structures *has* been demonstrated. Self-replicating chemical systems (Ful-

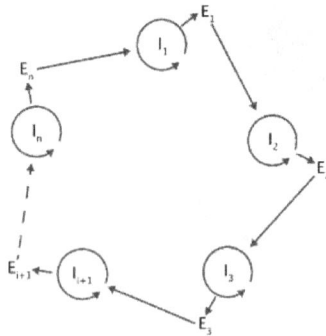

Fig. 35. The Hypercycle. Reproduced from Eigen & Schuster (1979). $E_i$ = an enzyme and $I_i$ = an RNA matrix. Each enzyme increases the replication rate of its subsequent RNA.

vene, Rotaxone) are well-known to scientists, and Eigen has explored autocatalytic sets of reactions that may be responsible for the self-organization of life (Eigen, 1971). His hypercycle (a reinforcing feedback loop) describes a closed loop of self-replicating molecules, each one of which catalyzes the creation of its successor (see Figure 35). Eigen successfully demonstrated that this chemical cycle displays Darwinian natural selection at the molecular level (Bahnzhaf, 2010.)

There is a plethora of underlying mechanisms and feedback loops responsible for DNA formation and function, operating in a symphony of self-organization. DNA replication is started by an initiator protein (DnaA) which seeks out those inter-base sites where the hydrogen bonds are weakest to start the cleaving of the double helix. (There are also inhibitor proteins to stop this activity.) This is followed by successive waves of Adenosine Tri-Phosphate (ATP) binding, hydrolysis, and release (acting in a feedback loop) causing the helicase enzyme to slide along the DNA molecule and unzip it. Subsequently, hydrogen bonding causes free bases in solution to bind to the single DNA strand, but adenine binds only with thymine and guanine binds only with cytosine due to the physical structure of those molecules. They fit together in a "lock-and-key" mechanism as shown in figure 36. In this feedback loop, hydrogen bonding is the attractive force while the physical shapes deter bonding.

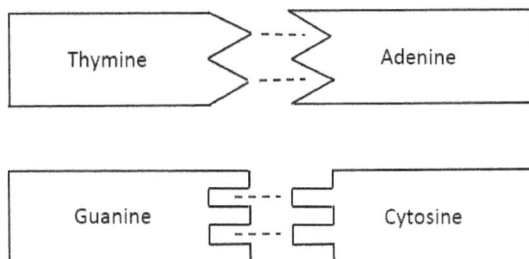

Figure 36. The "Lock-and-Key" Mechanism for Base Pair Bonding in DNA

There are many feedback loops associated with DNA. Systems Thinking helps explain the self-organized structures and underlying physical, electrical, and chemical forces that yield DNA's structure and function.

## Discussion

Natural patterns confront us every day. Many seem to defy understanding. However, if one realizes that patterns represent physical, temporal, chemical, or emotional oscillations, it becomes easier to fathom them.

There have been previous attempts to explain natural patterns. D'Arcy Wentworth Thompson (1917) argued that physical laws and mechanics play a vital role in determining the structure and morphology of organisms. His main thrusts, however, seem to be to try to explain the evolution of living organisms using mathematics and the morphology of distortion; he never explains animal behavior, non-biological morphologies, or the feedback loops that yield oscillations and patterns.

273

In his seminal paper, Alan Turing (1952) developed the reaction-diffusion mechanism to explain patterns in living organisms based on chemical and physical laws. He states, "The theory does not make any new hypotheses; it merely suggests that certain well-known physical laws are sufficient to account for many of the facts." He comes close to the systems thinking philosophy (and indeed articulates several elements of it) but doesn't integrate them into a broadly-applicable theory; specifically he does not explain inanimate or behavior patterns.

Mano (2004) does a nice job explaining natural patterns as a result of self-organization, as well as the advantages of self-organization, but he does not invoke feedback loops or detail the biological, chemical, and physical mechanisms that yield self-organization. He also suggests that the patterns we see were "stumbled upon" by chance. He proposes self-organization as a solution to a biological problem: "One of the mysteries of biology is how the enormous amount of morphogenic, physiological and behavioral complexity of living organisms can be achieved with the limited amount of genetic information available within the genome. Self-organization is one solution to this problem............ Through self-organization, evolution has stumbled upon a wide range of extremely efficient, relatively simple solutions for solving very complex problems."

In his ground-breaking book *The Life of the Cosmos*, Lee Smolin (1999) hypothesizes that our entire universe is a self-organized system, rife with feedback loops, hierarchies, and emergent patterns at all scales in galaxies, planetary systems, and atoms. He speculates that the laws of physics (and their concomitant universes) are not fixed, but evolve over time in a form of Darwinian natural selection. Although the book is full of systems thinking concepts, Smolin does not integrate them into an overarching systems thinking framework.

In his beautiful book, *Patterns in Nature*, Philip Ball (2016) describes dozens of natural patterns, including mud cracks, fern spirals, fractals, animal skin patterns, beehives, herding/schooling patterns, and flow patterns. He attempts to explain each using scientific principles, and he even refers to feedback loops occasionally. But he doesn't articulate the underlying causality in all these and misses integrating them into an overarching systems theory explanation involving causal loops, feedback, and hierarchies.

Parveen (2017) explains patterns mathematically, arguing that Fibonacci sequences often provide the most efficient light-gathering structures or the highest surface area: volume ratios, but he does not explain how natural systems generate those mathematical structures. Although these explanations have merit, they miss the integrated theory afforded by Systems Thinking, specifically the theory that emergent patterns represent oscillations caused by self-organized structure (feedback loops) which in turn, are caused by underlying forces, and that these apply to both living and non-living entities as well as to physical, temporal, emotional, and behavioral patterns. These concepts are the essence of Systems Thinking.

Systems Thinking addresses many questions about natural patterns by invoking concepts such as feedback loops, self-organization, hierarchies, and emergence. However, it also introduces others.

The Nature of the Universe
- As Lee Smolin (1999) suggests, is our whole universe a complex self-organized system, and can the universe exist and be self-organized without a prime mover?
- What are the minimum requirements for a universe to self-organize?
- Are the natural laws such as gravity really laws or are they just properties of the space we inhabit?

- Is life itself an emergent property deriving from the self-organization of hierarchical elements?
- Should our definition of "life" not be binary (either living or not-living) but more a continuum with some threshold beyond which we define as "alive"?
- Is self-awareness an emergent property of a sufficiently complex self-organized brain, and can any sufficiently complex entity become self-aware?
- What is the next step of emergence for humanity--the inter-connectedness of all humans into 1 super-being (Monat, 2017)?

These intriguing questions exhort study through the application of Systems Thinking.

**Conclusion**

In this paper, we have explored some of the patterns observed in nature, and used Systems Thinking to explain how and why they came to be. The concepts of feedback loops, hierarchies, self-organization, and emergence contribute to our understanding of these natural patterns. This understanding may help us to influence our natural world in positive ways, and to facilitate the design and improvement of human-designed systems, which is a topic for a future paper.

# REFERENCES

Bahnzhaf, W., "Self-Organizing Systems," In R. Meyer (Ed.), *Encyclopedia of Complexity and Systems Science*, Springer, 2010, 8040-8050

Ball, Philip, *Patterns in Nature: Why the Natural World Looks the Way It Does*, University of Chicago Press, 2016

Beckenkamp, Martin, "The Herd Moves? Emergence and Self-Organization in Collective Actors," Max Planck Institute for Research on Collective Goods, Bonn, 2006

Camazine, Scott, Jean-Louis Deneubourg, Nigel R. Franks, James Sneyd, Guy Theraulaz, & Eric Bonabeau, *Self-Organization in Biological Systems*, Princeton University Press, 2001.

Deneubourg, Jean-Louis, S. Aron, S. Goss, and J. M. Pasteels, "The Self-Organizing Exploratory Pattern of the Argentine Ant," *Journal of Insect Behavior* **3**:2, 1990

Deneubourg, J. L., & S. Goss, "Collective Patterns and Decision-Making," *Ethology, Ecology & Evolution*, **1**, 1989, 295-311.

Eigen, Manfred, "Self-Organization of Matter and the Evolution of Biological Macromolecules," *Die Naturwissenschaften (The Science of Nature,)* **58**:10, 1971, 465–523

Eigen M. and P. Schuster. *The Hypercycle: A Principle of Natural Self-Organization*, Springer, Berlin, 1979

Forterre P, Filée J, Myllykallio H., "Origin and Evolution of DNA and DNA Replication Machineries," In: *Madame Curie Bioscience Database*, Landes Bioscience; Austin,TX, 2000-2013

Fratzl, Peter, and Weinkamer, Richard, "Nature's Hierarchical Materials," *Progress in Materials Science*, **52**:8, 2007, 1263-1334

Gerola, H. & Seiden, P. E., "Stochastic Star Formation and Spiral Structure of Galaxies," *Astrophysical Journal, Part 1*, **223**, 1978, 129

Howard Hughes Medical Institute, 1/25/2005, http://www.hhmi.org/askascientist/answers/what_is_the_molecular_mechanism_for_stripes_in_zebras.html, accessed 8 August 2011

Hoyle, Sir Fred, and Wickramasinghe, Chandra, *Evolution from Space*, Simon & Schuster, NY, 1981

Johnson, Stephen, *EMERGENCE: The Connected Lives of Ants, Brains, Cities, and Software"* Scribner, 2002

Keller, M. A., Turchyn, A. V., and Ralser, M., "Non-Enzymatic Glycolysis and Pentose Phosphate Pathway-Like Reactions in a Plausible Archean Ocean," *Molecular Systems Biology*, **10**:725, 2014

Makse, Hernan, "Stratification in Large-Scale Aeolian Dunes," http://www-levich.engr.ccny.cuny.edu/~hmakse/rocks.html, accessed 4 December 2017

Mano, Jean-Pierre, "Self-Organization in Natural Systems," Institut de Recherche de l'informatique de Toulouse, TFG-AOSE 01.07.04 Roma, 2004. Accessed 7/13/2011.

Meadows, Dana, *Thinking in Systems*, Chelsea Green Publishing, White River Junction, VT, 2008, 83

Milgrom, Mordehai, "MOND Theory," *Canadian Journal of Physics*, **93**:2, 2015, 107

Miller, Noam Y., and Robert Gerlai. "Oscillations in Shoal Cohesion in Zebrafish (Danio Rerio)," *Behavioural Brain Research*, **193**:1, 2008, 148–151

Miller, Stanley L. (1953). "Production of Amino Acids Under Possible Primitive Earth Conditions," *Science*, **117** (3046): 528–9.

Miller, Stanley L.; Harold C. Urey (1959). "Organic Compound Synthesis on the Primitive Earth," *Science*, **130** (3370): 245–51.

Monat, J. P., "The Emergence of Humanity's Self-Awareness," *Futures Journal*, **86**, 2017, 27-35

Monat, J. P., and Gannon, T.F., "What Is Systems Thinking? A Review of Selected Literature Plus Recommendations," *Am. J. of Systems Science*, **4**:2, 2015

Monat, J. P., and Gannon, T.F., "Using Systems Thinking to Analyze ISIS," *American Journal of Systems Science*, **4**:2, 2015, 36-49

Monat, J. P., and Gannon, T.F., *Using Systems Thinking to Solve Real-World Problems*, College Publications, London, 2017

Muhammad Saif Ullah, Khalid; Imran Akhtar; and Haibo Dong, "Hydrodynamics of a Tandem Fish School with Asynchronous Undulation of Individuals," *Journal of Fluids and Structures*, **66**, 2016, 19-35

Myers, P. Z., "Chance and Regularity in the Development of the Fly Eye," http://scienceblogs.com/pharyngula/2006/03/chance_and_regularity_in_the_d.php

Oro, Joan, "Synthesis of Adenine from Ammonium Cyanide," *Biochemical and Biophysical Research Communications*, **2**:6, 1960, 407-412

Oro, Joan, "Synthesis of Purines Under Possible Primitive Earth Conditions. I. Adenine from Hydrogen Cyanide," *Archives of Biochemistry and Biophysics*, **94**:2, 1961, 217-227

Parveen, Nikhat, "Fibonacci in Nature," University of Georgia Dept. of Mathematics and Science Education, http://jwilson.coe.uga.edu/emat6680/parveen/fib_nature.htm, accessed 13 December 2017

Reynolds, C. W. (1987,) "Flocks, Herds, and Schools: A Distributed Behavioral Model," in *Computer Graphics*, **21**:4 (SIGGRAPH '87 Conference Proceedings) pages 25-34.

Reynolds, Craig, "Boids," https://www.red3d.com/cwr/boids/, accessed 29 November 2017

Scitable by Nature Education, "The Origin of Mitochondria and Chloroplasts," https://www.nature.com/scitable/content/the-origin-of-mitochondria-and-chloroplasts-14747702, accessed 10 December 2017

Smolin, Lee, *The Life of the Cosmos*, Oxford University Press, 1999

Smolin, Lee, "The Self-Organization of Space and Time," *Philosophical Transactions of the Royal Society A*, The Royal Society, 2003

Smyth, William D. and Moum, James N., "Ocean Mixing by Kelvin-Helmholtz Instability," *Oceanography*, **25**:2, 2012, 140-149

Theraulaz, G., Bonabeau, E., Deneubourg, J.L., "The Origin of Nest Complexity in Social Insects: Learning from the models of nest construction," *Complexity* **3**:6, 1998, 15-25

Thompson, D'Arcy Wentworth, *On Growth and Form*, Cambridge University Press, 1917

Turing, Alan, ""The Chemical Basis of Morphogenesis," *Philosophical Transactions of the Royal Society of London. Series B, Biological Sciences,* **237**:641, 1952, 37-72.

www.ingramcontent.com/pod-product-compliance
Lightning Source LLC
Chambersburg PA
CBHW060331200326
41519CB00011BA/1903